CAMBRIDGE LATIN AMERICAN STUDIES

29

DROUGHT AND IRRIGATION IN NORTH-EAST BRAZIL

THE SERIES

1 Simon Collier. *Ideas and Politics of Chilean Independence, 1808–1833*
2 Michael P. Costeloe. *Church Wealth in Mexico: A study of the Juzgado de Capellanías in the Archbishopric of Mexico, 1800–1856*
3 Peter Calvert. *The Mexican Revolution, 1910–1914: The Diplomacy of Anglo-American Conflict*
4 Richard Graham. *Britain and the Onset of Modernization in Brazil, 1850–1914*
5 Herbert S. Klein. *Parties and Political Change in Bolivia, 1880–1952*
6 Leslie Bethell. *The Abolition of the Brazilian Slave Trade: Britain, Brazil and the Slave Trade Question, 1807–1869*
7 David Barkin and Timothy King. *Regional Economic Development: The River Basin Approach in Mexico*
8 Celso Furtado. *Economic Development of Latin America: Historical Background and Contemporary Problems* (second edition)
9 William Paul McGreevey. *An Economic History of Colombia, 1845–1930*
10 D.A. Brading. *Miners and Merchants in Bourbon Mexico, 1763–1810*
11 Jan Bazant. *Alienation of Church Wealth in Mexico: Social and Economic Aspects of the Liberal Revolution, 1856–1875*
12 Brian R. Hamnett. *Politics and Trade in Southern Mexico, 1750–1821*
13 J. Valerie Fifer. *Bolivia: Land, Location, and Politics since 1825*
14 Peter Gerhard. *A Guide to the Historical Geography of New Spain*
15 P.J. Bakewell. *Silver Mining and Society in Colonial Mexico, Zacatecas 1546–1700*
16 Kenneth R. Maxwell. *Conflicts and Conspiracies: Brazil and Portugal, 1750–1808*
17 Verena Martinez-Alier. *Marriage, Class and Colour in Nineteenth-Century Cuba: A Study of Racial Attitudes and Sexual Values in a Slave Society*
18 Tulio Halperín-Donghi. *Politics, Economics and Society in Argentina in the Revolutionary Period*
19 David Rock. *Politics in Argentina 1890–1930: the Rise and Fall of Radicalism*
20 Mario Góngora. *Studies in the Colonial History of Spanish America*
21 Arnold J. Bauer. *Chilean Rural Society from the Spanish Conquest to 1930*
22 James Lockhart and Enrique Otte. *Letters and People of the Spanish Indies: The Sixteenth Century*
23 Leslie B. Rout Jr. *The History of the African in Spanish America: from 1502 to the Present Day*
24 Jean A. Meyer. *The Cristero Rebellion: The Mexican People between Church and State, 1926–1929*
25 Stefan De Vylder. *Allende's Chile: The Political Economy of the Rise and Fall of the Unidad Popular*
26 Kenneth Duncan and Ian Rutledge with the collaboration of Colin Harding. *Land and Labour in Latin America: Essays on the Development of Agrarian Capitalism in the Nineteenth and Twentieth Centuries*
27 Guillermo Lora. *A History of the Bolivian Movement 1848–1971* edited and abridged by Laurence Whitehead and translated by Christine Whitehead
28 Victor Nunes Leal. *Coronelismo: The Municipality and Representative Government in Brazil.* Translated by June Henfrey

Drought and Irrigation in North-East Brazil

ANTHONY L. HALL

CAMBRIDGE UNIVERSITY PRESS
Cambridge
London New York Melbourne

Published by the Syndics of the Cambridge University Press
The Pitt Building, Trumpington Street, Cambridge CB2 1RP
Bentley House, 200 Euston Road, London NW1 2DB
32 East 57th Street, New York, NY 10022, USA
296 Beaconsfield Parade, Middle Park, Melbourne 3206, Australia

First published 1978

Printed in Great Britain
at the University Press, Cambridge

Library of Congress Cataloguing in Publication Data

Hall, Anthony L. 1947

Drought and irrigation in North-East Brazil.

(Cambridge Latin American studies; 29)

Originally presented as the author's thesis,
University of Glasgow, 1976.

Bibliography: p.

Includes index.

1. Agriculture – Economic aspects – Brazil, North-East.
2. Drought – Brazil, North-East. 3. Irrigation –
Brazil, North-East. I. Title. II. Series.
HD1872.H34 1978 338.1′0981′3 77-82497

ISBN 0 521 21811 X

Contents

Tables

To my parents,
Leslie and Giuliana
To Rejane
In memory of my grandparents,
Pietro and Gemma

Acknowledgements

This book is based on fieldwork carried out in North-East Brazil from September 1974 until November 1975. It was originally presented as a PhD thesis to the University of Glasgow in May 1976. I should like to express my gratitude to the Ministry of Overseas Development, whose research award made this field trip possible. My thanks are also due to the *Instituto Joaquim Nabuco de Pesquisas Sociais*, Recife, for their support, and in particular to Clovis Cavalcanti and his staff in the Department of Economics. I would also like to thank the Carnegie Trust and the Anglo-Brazilian Society for their financial help. A two-year research scholarship from the University of Glasgow, to whom my thanks are also due, enabled me to prepare and write up the original research. A special vote of thanks is reserved for Simon Mitchell of the Institute of Latin American Studies, Glasgow University, whose consistent help and encouragement made the research task far less daunting. Lastly, but by no means least, I wish to thank Professor Werner Baer of the University of Illinois for his assistance and constructive comments.

Anthony Hall

Recife, March 1977

Abbreviations

ASMIC	*Association pour l'Organisation de Missions de Coopération Technique*
BB	*Banco do Brasil*
BNB	*Banco do Nordeste do Brasil*
CHESF	*Companhia Hidroeléctrica do Vale do São Francisco*
CIDA	*Companhia Integrada de Desenvolvimento Agrícola*
CODEVASF	*Companhia de Desenvolvimento do Vale do São Francisco*
DNOCS	*Departamento Nacional de Obras Contra as Sêcas*
DNOS	*Departamento Nacional de Obras de Saneamento*
FIPE	*Fundação Instituto de Pesquisas Econômicas*
FUNRURAL	*Fundo de Assistência para o Trabalhador Rural*
GEIDA	*Grupo Executivo de Irrigação para o Desenvolvimento Agrícola*
GTDN	*Grupo de Trabalho para o Desenvolvimento do Nordeste*
GVJ	*Grupo de Estudo de Base do Vale do Jaguaribe*
IBGE	*Instituto Brasileiro de Geografia e Estatística*
IBRD	International Bank for Reconstruction and Development
IFOCS	*Inspectoria Federal de Obras Contra as Sêcas*
IJNPS	*Instituto Joaquim Nabuco de Pesquisas Sociais*
INCRA	*Instituto Brasileiro de Colonização e Reforma Agrária*
IOCS	*Inspectoria de Obras Contra as Sêcas*
IPEA	*Instituto de Planejamento Econômico e Social*
IRYDA	*Instituto Nacional de Reforma y Desarrollo Agrario*
MINTER	*Ministerio do Interior*
POLAMAZONIA	*Programa de Polos Agropecuários e Agrominerais da Amazônia*
POLONORDESTE	*Programa de Desenvolvimento de Áreas Integradas do Nordeste*
PROTERRA	*Programa de Redistribuição de Terras e Estímulos à Agro-Industria do Nordeste*
PPI	*Programa Plurianual de Irrigação*
SCET	*Societé pour l'Equipment du Territoire-Coopération*

SEPLAN	*Secretaria de Planejamento*
SIRAC	*Serviços Integrados de Assesoria e Consultoria*
SUDENE	*Superintendência do Desenvolvimento do Nordeste*
SUVALE	*Superintendência do Vale do São Francisco*
USAID	United States Agency for International Aid and Development

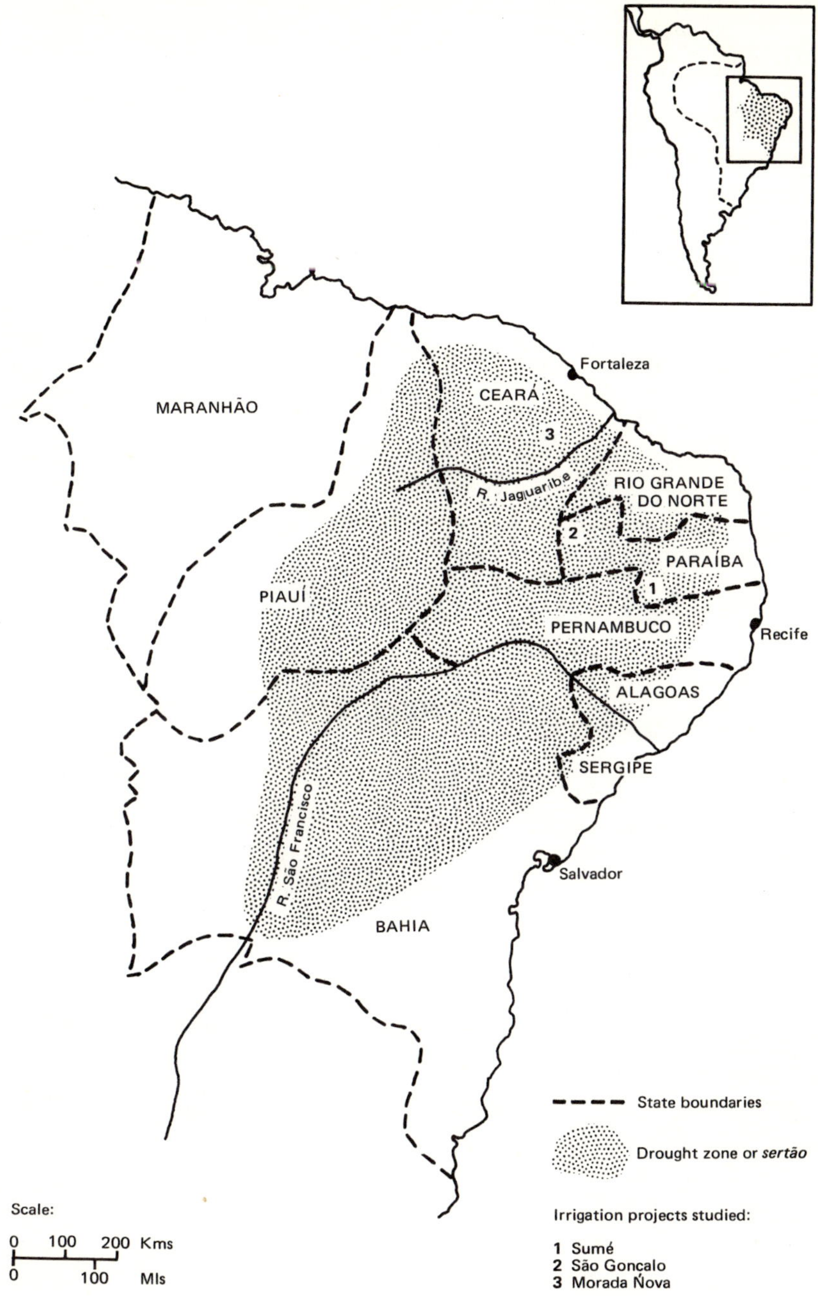

The drought zone of North-East Brazil

1. Drought in North-East Brazil – a general picture

Drought in North-East Brazil has given rise to considerable human suffering for at least two hundred years, but only during this century has the Brazilian government tried to tackle the problem. Drought effects are manifested in the form of widespread rural unemployment, poverty, starvation and subsequent migration from affected areas. Such characteristics, while to some extent a permanent feature of rural society in the *sertão*, the interior of the North-East, become seriously accentuated during prolonged periods of scarce rainfall. Such periods are inevitably labelled 'droughts' and occur in their most extreme form about once every century, the last catastrophic drought having struck in 1877–9. Milder periods of sparse rainfall occur at more intermittent intervals, the last being in 1970 and 1976. This book examines the efficacy of irrigation schemes, the latest government strategy designed to cope with the 'drought problem'.

The government agency responsible for fighting the drought, the *Departamento Nacional de Obras Contra as Sêcas* (DNOCS), experimented with irrigation in a very small way as far back as the 1930s, but until recently no systematic attempt had been made to expand it as a development strategy capable of tackling the drought problems on a broad scale. Rather, as will be shown in greater detail below, efforts were concentrated on building reservoirs (*açudes*), boring wells and constructing roads. The first broad study of the potential advantages of irrigation in the North-East was produced in 1968 by the newly created *Grupo Executivo de Irrigação para o Desenvolvimento Agrícola* (GEIDA), in conjunction with an Israeli firm of consultants (Tahal).

Out of this preliminary analysis GEIDA published the *Programa Plurianual de Irrigação* (*PPI*) in 1971, which lays out comprehensive plans for irrigation in the North-East as part of a national strategy.[1] It also forms part of the government's overall development plan for the country under the Plan for National Integration (PIN), brought to the statute books in 1970. Within the terms of the *PPI*, DNOCS was entrusted with the responsibility for irrigation projects falling within a large part of the 'drought polygon', the semi-arid region which covers almost one million square kilometres from the state of Piauí in the north to Minas Gerais in the south, and which has a rural population of some eighteen million. The *Superintendência do Vale do São Francisco* (SUVALE), recently renamed as the

Companhia de Desenvolvimento do Vale do São Francisco (CODEVASF), was charged with developing irrigation along the valley of the river São Francisco, the largest in the North-East.

By 1976 DNOCS had twelve projects in operation, having implemented schemes rather more quickly than CODEVASF, which had only two projects functioning at that time. In this study, therefore, I have concentrated on the DNOCS irrigation programme, where the greater volume offers a more realistic basis from which to draw conclusions concerning the efficacy of irrigation.

DNOCS has three basic objectives behind its irrigation programme: (1) to stabilize the rural population by providing employment to families which would otherwise be forced to migrate because of drought, (2) to increase family incomes, thus attacking the problem of rural poverty, and (3) to increase food supplies for the North-Eastern market.[2] Using general documentary information as well as primary data collected from three case studies of DNOCS projects, I have examined whether or not these objectives are being met and whether the 'drought problem' is being effectively tackled.

The recorded history of droughts in the North-East goes back to the sixteenth century, when Father Fernão Cardim described in 1583 how indigenous tribes migrated to the coast when rainfall was scarce. He estimated that during the 1587 drought, for example, 5,000 Cariris and Tabajaras fled the interior.[3] However, the human impact of the drought was minimized by the fact that, until the late eighteenth and early nineteenth centuries, the *sertão* held little attraction for the Portuguese colonizers. Its harsh climate and relatively infertile soils, when compared with the rich land along the humid Atlantic coastline, as well as resistance from the indians, discouraged early penetration and settlement. The Dutch appear to have made some journeys into the interior during the mid-seventeenth century, notably along the valleys of the rivers Acaraú and Jaguaribe.[4] However, they did not venture too far into the *sertão* but were content, as one observer put it, 'to stay on the coast, barely penetrating a few leagues'.[5]

The drive to settle the interior and establish cattle estates or *fazendas* received its main push during the late seventeenth century after the Dutch had been expelled. Land grants were made by the Portuguese Crown to colonizers who travelled inland and subdued the indians. Several of the early explorers came from the south of Brazil in order to make their fortunes in the North-East: early *bandeirantes* such as Domingos de Freitas de Azevedo who penetrated the São Francisco valley, and Domingos Jorge Velho, Matias Cardoso de Almeida and Morais Navarro who conquered the indians along the rivers Açú and Jaguaribe. The main valleys taken over, other areas of the interior were quickly gained by the end of the eighteenth century.[6] Even by the early 1800s cattle *fazendas* had become well established in many parts of the *sertão*, some, in Pernambuco for example, branding up to 500 head per year.[7]

During the early colonial period, then, individual families came to acquire estates which often covered enormous areas, up to hundreds or even thousands of square kilometres.[8] This division of land into large, private properties in rural areas was to play a major role in determining the concentration of economic and political power in the hands of *latifundistas* and, as we shall see, had important repercussions on the nature and efficacy of future drought relief measures.

These estates, which supplied the more densely populated, sugar-producing coastal zone, the *zona da mata*, supported very few people themselves. The landowner needed only a small number of *vaqueiros* and *moradores* to look after his cattle. By the end of the eighteenth century the *sertão* population consisted of landowners, the indians and a few *mestiço* labourers. The indigenous tribes often raided the estates when drought drove away the game. They were also defending their major sources of fish and game, the fertile river valleys and more humid uplands (*serras*), which were the first areas to be colonized in the interior.

During the colonial period, the human impact of the drought was reduced by the fact that the population at risk was rather small. It has been estimated that the *sertão* had a population of 726,000 in 1800, although clearly such evidence must be treated with caution.[9] In addition, cattle estates were relatively well protected from the drought by virtue of being situated on river margins which provided pasture during the dry season. Many landowners also moved their cattle to graze in the cooler and more humid *serras* such as Baturité, Ibiapaba and Araripe. Nevertheless, the primitive methods used by the colonizers, as well as the lack of concern for providing an adequate system of food supply and storage for what population there was, led to the death of many cattle and reports of quite severe human suffering. One contemporary observer noted that in 1792 the drought in Ceará killed 'thousands',[10] while another reported a death roll of 30,000 during the drought of 1825.[11] The scarcity of food supplies resulted in manioc flour being sold at ten times its normal price during drought periods.[12]

Up to the mid-nineteenth century the North-East had seen a dozen serious droughts, but it was the human calamity of 1877–9 which eventually led to government action being taken for the first time, although somewhat belatedly. A combination of hunger and disease resulting from migration and the herding together of refugees (known as *retirantes* or *flagelados*) into camps is reputed to have claimed 500,000 lives during this period.[13] This was far more serious than any drought which had been experienced before, and was due chiefly to the fact that since the mid-nineteenth century the population at risk had increased considerably. Until then the drought impact had been limited because the cattle economy of the *sertão* had absorbed very little labour. However, this picture was radically changed from 1850 onwards. Aided by a total absence of droughts from 1845–76, a drought-resisting type of arboreal cotton (*mocó*) was introduced into the interior. Cotton, of course, required much more labour than cattle-raising and, stimulated by the cotton shortage resulting from the American Civil War, the crop prospered and attracted large numbers of rural labourers from the neighbouring *agreste* region. A mixed cattle and cotton economy was thus set up whose elements complemented each other rather well. Amongst other advantages, cotton seed cake provided fodder for cattle. A cotton boom took place in which other crops were sacrificed and rapid profits were made. Exports of cotton from Ceará increased sharply from 165,265 kilos in 1845 to 7,906,944 kilos in 1871.[14]

In order to feed themselves, the rural work-force cultivated subsistence crops (maize, beans and manioc) in conjunction with, and frequently on a similar sharecropping basis to, arrangements regarding cotton. However, although adequate in years of normal rainfall, during a drought subsistence crops failed, removing the only source of food for the labouring population. The economic situation was worsened by the drop in cotton prices after 1871, decreasing both profits and wages.[15] The crop failure caused by the 1877–9 drought combined with a precarious economic situation so that, unable even to feed themselves, huge numbers of rural inhabitants were forced to migrate in search of relief elsewhere. Landowners were either unwilling or unable to shoulder the economic burden of supporting their employees during this difficult period.[16]

For the first time, then, a drought was documented in graphic detail and the vulnerability of the low-income rural mass became tragically apparent. As the winter rains failed to arrive and their stocks of food dwindled rapidly, the rural workers of the *sertão* and their families started migrating to the humid *serras*, to the rubber boom in the Amazon and to the towns. The population of Ceará's state capital, Fortaleza, was swollen from its normal 40,000 in 1877 to 160,000 in 1879 by incoming *flagelados*. The poor conditions which awaited them and their herding together into camps which, by 1915, had become known as 'concentration camps', caused 60,000 deaths from hunger and disease (cholera and smallpox) in this city alone.[17]

Crimes were committed both by starving refugees struggling to survive and by citizens undertaking reprisals against the unwelcome newcomers. The latter's crimes, Theóphilo notes, 'often went unpunished because the *retirante* was considered a leprous dog who was going to stain the land'.[18] *Retirantes* flocked to the city, only to be sadly disillusioned with the lack of provision made by the government for their care. Theóphilo observed: 'The sad procession paraded along the streets of the capital at all hours . . . Real animated skeletons, with skin blackened by the dust from the roads and stuck to their bones, held out their hands begging from everyone they met.'[19]

However, although for the first time in 1877 a drought had come to be considered officially as a national disaster, relief was not plentiful and came in the form of short-term food supplies. Its effectiveness was reduced further still by corruption amongst officials in charge of distributing aid to refugees.[20] Theóphilo complained bitterly about the lack of government relief, not just during the 1877 drought, but during those of 1900 and 1915 also. A large proportion of what official aid was forthcoming seems to have been spent on subsidizing emigrants' sea passages rather than on helping those who had no option but to stay.[21]

During the three decades following the 1877 drought the government made a slow start at devising projects which, it hoped, would lessen the drought impact. In December 1877 a National Commission of Enquiry was set up which recommended the construction of reservoirs (*açudes*), the improvement of communications by building railways and roads and by improving the harbours, as well as better water control to be achieved through digging wells and the building of a canal to link the rivers Jaguaribe and São Francisco. For a variety of reasons, however, including technical difficulties, poor administration, lack of funds and sheer lack of government interest, most of these proposals did not get beyond the planning stage.

One project which was completed was the Açude Cedro (later renamed Quixadá), constructed by the French engineer Jules Revy between 1884 and 1906. It was, however, criticized at the time by President-elect Afonso Pena as 'pretty but useless'.[22] Despite this observation, reservoir-building or *açudagem* was to set the pattern for drought relief works over the next sixty years: stored water which benefitted large landowners and protected their cattle by providing pasture and watering facilities but which, although affording some protection to a handful of adjacent smallholders, left most of the low-income agricultural population untouched. Three technical missions were also established from 1904–6 to deal with public works in Ceará and Rio Grande do Norte, but achieved very little.[23]

Increasing dissatisfaction with the lack of action taken against the drought and with the limited success of what measures had been implemented led to the creation of the *Superintendência dos Estudos e Obras Contra os Efeitos da Sêca* in 1906, renamed two years later as the *Inspectoria de Obras Contra as Sêcas* (IOCS) and again as the *Inspectoria Federal de Obras Contra as Sêcas* (IFOCS) in 1918, with its headquarters in Rio de Janeiro. Dominated by natural scientists and civil engineers, IFOCS believed that the most effective way of fighting the drought was to build dams in river valleys and accumulate water in *açudes*. This was its major emphasis but, apart from reservoir-building, IFOCS also started to bore wells, give subsidies to private landowners for building their own small reservoirs, set up meteorological stations, carry out geological surveys and build roads. Highway construction was seen then, as it still is today, not only as a means of employing labour, but also as a way of improving communications in order to facilitate the rapid evacuation of the drought-affected population and the faster shipment of relief supplies.

Several factors have influenced the flow of federal funds to finance drought relief works in the North-East. Of primary importance has been the advent of a drought itself, when public alarm is at its peak and the urgency of the need is brought home more vividly. During the aftermath of a drought there has inevitably been a tailing-off period when both interest and funds seem to decrease. The flow of funds to IFOCS was not only influenced by the immediacy of the drought but by the general economic situation and by how much influence North-Eastern politicians had at federal level. Consequently, there were enormous fluctuations in the flow of federal money. Under the administration of Epitacio Pessoa, a native of the North-Eastern state of Paraíba, the amount of drought relief rose from four million *milreis* in 1916–19 to 142 million *milreis* in 1921–2, financed by borrowing from national banks and from abroad (including US$75 million raised in New York). American and British engineers were brought in and many new projects were started.[24] These included, as one contemporary English observer noted, the building of 1,400 kilometres of roads, 196 public and private reservoirs, numerous wells and 500 kilometres of railway.[25] However, many of these projects received setbacks during the subsequent government of Artur Bernardes (a native of the relatively drought-free state of Minas Gerais) at a time of financial crisis.

The serious drought of 1930–2 stimulated relief work and, with *paraibano* José Américo de Almeida as his Minister of Public Works, President Getulio Vargas increased spending on the North-East to 10% of federal revenue in 1932. In 1934 under the new Constitution 4% of total tax revenue was set aside for drought relief.[26] After the 1945 drought IFOCS was given its present name, the *Departamento Nacional de Obras Contra as Sêcas* (DNOCS), and its headquarters was transferred from Rio to Fortaleza. The onset of a drought, however, still meant that short-term emergency relief had to be supplied in order to avoid mass starvation. In the 1932 crisis, for example, IFOCS employed 220,000 *flagelados* building reservoirs and roads.[27] *Açudagem* continued to be the main longer-term anti-drought strategy, so that the volume of water stored in publicly owned reservoirs rose from 2.9 billion cubic metres in 1955 to 7.8 billion in 1960,[28] reaching the figure of 11.4 billion cubic metres (in 253 reservoirs) by 1972.[29] In 1945 the *Companhia Hidroeléctrica do Vale do São Francisco* (CHESF) was set up to develop hydroelectric resources along the river São Francisco. Three years later the *Commissão do Vale do São Francisco* (CVSF) was established to expand the economic and social infrastructure of the valley.

During the 1950s it became increasingly clear that DNOCS's policies were doing very little to provide long-term protection for the mass of the rural population in the North-Eastern interior. As the above figures show, however, the agency was not to be discouraged from its reservoir-building strategy. Nevertheless, the fact that it was necessary to provide relief for 60,000 *flagelados* during the 1951 drought underlined the need for new thinking and for broader socio-economic policies which would have a wider impact than mere reservoir-building. As a first step the *Banco do Nordeste do Brasil* (BNB) was set up in 1952 with funds put aside by the 1946 Constitution. The BNB was, however, a commercial bank rather than a development bank and gave credit to farmers and industrialists on that basis. Nevertheless, its research division ETENE (*Departamento de Estudos Econômicos do Nordeste*) examined social and economic problems of the region and set a precedent in this respect. Preoccupation with examining regional problems in depth also led to the formation of a special study group, the *Grupo de Trabalho para o Desenvolvimento do Nordeste* (GTDN) in 1956.

It was the calamitous 1958 drought which finally discredited the *política hidráulica* of DNOCS and set in motion a series of events which eventually resulted in the realignment of long-term drought relief strategy away from reservoir construction and towards irrigation. When the drought struck, the North-East had accumulated 6.7 billion cubic metres of water in over 200 public *açudes*, not counting the thousands of small, private reservoirs belonging to landowners. The region had three times the national average road mileage, commercial credit was available to support private investment and the electricity-generating capacity had doubled upon completion of the Paulo Affonso falls scheme in 1956.

Nevertheless, none of these infrastructural improvements prevented the number of refugees employed on work fronts from reaching 536,000, or 13% of the total North-Eastern population, in 1958. At the height of the drought the total number of people affected reached 2¾ million, including dependents, or 21% of the 'drought polygon' population.[30] Thousands of *retirantes* were forced to leave the region in search of temporary employment in the Centre-South and coastal towns of the North-East. Profits were made by exploiting their hardship to provide cheap labour for the construction of Brasilia. The unfortunate emigrants or *candangos*, as they became known, were taken in droves on open trucks, the infamous *pau de araras*, often to be abandoned on the roadside by drivers anxious to avoid discovery by the police.[31]

Due to the severity of the drought, DNOCS was granted a record amount of funds by Congress for relief work, to be used principally on work fronts building dams and roads. Money was also provided to assist private landowners to build their own reservoirs, as well as for using stored water capacity to improve domestic supplies and electricity generation. President Kubitschek ordered construction to be resumed on the enormous Orós dam in Ceará which had been started under Epitacio Pessoa. Its four billion cubic metres would, at a stroke, double the amount of stored water in the North-East, forming a virtual inland sea and having the capacity to irrigate 100,000 hectares (1 hectare = 2.47 acres).

However, although to all intents and purposes DNOCS was riding on the crest of a wave, it soon became discredited by tales of bribery and corruption in the administration of relief funds, as well as by its dubious unofficial role in the 1958 state gubernatorial elections. DNOCS became closely associated with what were known as the 'drought industrialists', the *indústrias da sêca*, a term used to describe anyone who exploited the drought situation for his own profit. These included simple thieves of relief shipments, corrupt officials in charge of public works and those in search of quick profits caused by the food shortages. At higher levels political pressures and connections were often used to influence DNOCS officials in the siting of roads and dams.[32] Landowners bundled their labourers off to a work front when the drought struck so that the government and not they should support them during the unproductive period. This tactic also had the advantage of keeping cheap labour in the area, minimizing the risk that workers would emigrate permanently. Federal deputies used their influence to secure relief funds for their electoral zone, ensuring their clients' votes for the future.[33] It was not uncommon, either, for relief to be administered on an individual basis, receipt or non-receipt depending upon past voting habits. One critic condemned DNOCS as 'serving the politics of the North-East and not the North-East itself. Its function is to serve the important families of the region.'[34]

DNOCS's poor reputation was made worse by its role in the 1958 October elections in which several important state governorships were disputed. The ruling groups tried, although unsuccessfully as it turned out, to use the power that DNOCS held to influence voting through the selective distribution of relief to those who promised to vote for Socialist Party and Labour Party candidates. In addition, it was alleged that DNOCS employees were used to frighten voters, DNOCS trucks were used to transport voters in project areas, election officials were bought and votes were purchased.[35] By the late 1950s DNOCS was totally discredited, its association with the *indústrias da sêca* and electoral corruption known, and its *açudagem* philosophy demonstrably a failure.

In retrospect it is surprising in some ways that irrigation as an anti-drought tactic was not more widely adopted earlier on. As early as 1931 IFOCS had been legally entitled to expropriate lands downstream from dams for irrigation projects, although it does not seem that these powers were ever used.[36] It had, in fact, been realized very early on, in the days when the first anti-drought measures were being planned, that reservoir-building would have to be supplemented by irrigation if the benefits of water storage were to be more widely spread. In the 1890s engineers were ignoring the advice of irrigation experts in the North-East.[37] Arrojado Lisboa, director of IFOCS (then IOCS) from 1909–12, stressed the need for irrigation, as did José Augusto Trindade, first director of IFOCS's research and extension division.[38] Even in 1937 a report commissioned by IFOCS stressed most emphatically the need to utilize stored water for irrigation, 'Irrigation will solve the problem . . . Only in this way will it be possible to keep the *sertanejo* in the *sertão*.'[39] In 1950 the then director of DNOCS, Vinicius Berredo, restated the need for carefully controlled irrigation projects on the expropriated lands of the hydraulic basins of existing reservoirs.[40] At this time, however, the only land irrigated under the initiative of DNOCS was a few hundred hectares on five or six projects, the land in which belonged to private owners, and which dated from the 1930s.[41]

DNOCS was reluctant to expropriate lands for two basic reasons. Firstly, the organization was dominated by civil engineers who measured their success by additions to the amount of water stored in *açudes* made during their term of office. Secondly, the DNOCS bureaucracy had developed close links with local power holders and were careful not to antagonize any more large landowners than necessary. This was especially so since many large property owners had seen their lands covered by public reservoirs already. Official publications today, however, lay most of the blame for the slow progress with irrigation on the 'unreceptive' attitudes of landowners rather than on DNOCS's reluctance to compromise itself politically.[42]

The period immediately following the 1958 drought was, then, one of crisis caused both by the size of the human catastrophe, as well as by the political controversies surrounding DNOCS. In this atmosphere, a group of economists at the National Development Bank (BNDE) had, at the request of President Kubitschek, prepared an economic analysis of the North-East's problems. Under the leadership of Celso Furtado, the 'Furtado Report', as it became known, recommended a comprehensive development programme designed to redress the severe economic imbalance between the North-East and Centre-South, based on three strategies: the use of reservoir water for irrigation schemes, organized colonization of wetter areas such as Maranhão and more efficient use of the sugar-cane lands along the coast.[43]

The failure of existing organizations to cope with the North-East's problems, together with the new ideas put forward by Furtado's report in this crisis atmosphere, led President Kubitschek to set up a new body, the *Superintendência do Desenvolvimento do Nordeste* (SUDENE), in 1959. SUDENE was given unprecedented powers to coordinate the activities of existing agencies in the region, such as DNOCS, CVSF and the BNB, as well as to formulate policies of its own.

In order that SUDENE should have control over DNOCS's activities the latter agency had to be transferred from the Ministry of Building and Public Works to join SUDENE in the Ministry of the Interior (MINTER). This was in line with the changed emphasis away from pure dam and road building to a broader irrigation programme. DNOCS strongly opposed this move, to the extent of engaging in the attempted character assassination of SUDENE's director, Celso Furtado, with accusations that he was a Communist. However, with the support of the nine North-Eastern state governors on SUDENE's board of directors, the measure was passed.[44]

It was really the formation of SUDENE which therefore forced DNOCS to start seriously considering irrigation as an anti-drought strategy. Rhetoric would no longer be enough; practical results were expected. DNOCS had eventually come around to admitting the 'limited effects of large-scale *açudagem* as a source of employment for the population of the semi-arid zone'.[45] However, official publications of the period take pains to express DNOCS's considerable apprehension over irrigation as involving 'unique and complex problems'.[46] Nevertheless, the point was re-emphasized in

SUDENE's regional development plans so that there was really no getting away from the fact that, for the foreseeable future at least, irrigation as an official strategy was here to stay.[47]

In fact it was SUDENE and not DNOCS which took the initiative in starting serious study of irrigation potential. In conjunction with a French mission[48] SUDENE carried out a detailed study between 1962 and 1965 of the Jaguaribe valley, the second largest valley of the region and with a huge volume of water stored in reservoirs from the *açudagem* period. The research team recommended the establishment of irrigation projects on the basis of already existing reservoirs.[49] This led in 1968 to the start of work on the first publicly owned irrigation scheme in the North-East, the Morada Nova project, fed by the Barabuiú reservoir. It will be examined in greater detail below as one of the case studies.

During the same period, in 1967, the CVSF was renamed the *Superintendência do Vale do São Francisco* (SUVALE) and alloted 1% of federal tax revenues per annum. However, most of its activities were devoted towards improving access roads and minor port facilities as well as medical and veterinary services. Like its partner in water resources development, DNOCS, the agency had accomplished little in the way of longer-term projects such as river regularization or irrigation.

During the 1960s, then, there was some optimism that irrigation had the potential to provide protection against the drought where reservoir-building had failed. SUDENE's initial efforts and DNOCS's official change of strategy were followed up by the formation of GEIDA in 1968 to undertake a comprehensive study of irrigation potential in Brazil. The 1970 drought served as a grim reminder that the rural population of the *sertão*, or certain sectors of it at least, were still as vulnerable as ever to the drought. The late arrival or absence of winter rains from April to December of that year forced thousands of small farmers off the land in search of employment elsewhere. Dependent on traditional subsistence crops for their food supply, these groups saw a drastic reduction in the volume of production because of the drought. From 1969–70 in drought-affected areas, production of beans decreased by 78%, maize by 86% and manioc by 61%.[50] 62% of the 'drought polygon' was affected, throwing 35% of the labour force out of work. At the height of relief operations in October–November of 1970, 500,000 men were employed on 112 work fronts. This made some 3½ million people totally dependent on drought relief, that is 19% of the total North-Eastern rural population of 18 million.[51]

Apart from using *flagelado* labour to build roads and dams on work fronts (at a daily wage of two *cruzeiros*), food was distributed, medical assistance was given at special posts and accommodation was provided for the refugees. It is estimated that the total cost of the drought in terms of lost agricultural production and emergency funds was in the order of US$400 million.[52] The distribution of relief seems to have been relatively free of corruption compared with the 1958 drought, thanks to some extent perhaps to close army supervision. Nevertheless, one indication of how little times have changed lies in the reports of root poisoning suffered by the desperately hungry in 1970,[53] reminiscent of the graphic accounts given by Theóphilo a century earlier.[54] Similarly, the hardships imposed by the drought manifested themselves in ways which had long been familiar in the region – prostitution, banditry and other crimes of violence.[55]

The period during and immediately after the 1970 drought resounded with clashes between the North-Eastern state governors and the federal government. Although a substantial amount had been allocated for relief measures, some US$13 million, there was considerable disquiet expressed about the possibility of the North-East being 'abandoned' by the government. In June 1970, after a visit to drought-affected areas, President Medici made a much-quoted and emotional speech in the SUDENE auditorium in Recife in which he affirmed his commitment to the North-East: 'I want to say to the people of the North-East that I do not promise you anything. I do not promise miracles or transformations, nor money nor favours, nor do I solicit sacrifices or votes, nor organization of charity. I only say that everything has to begin to change.'[56]

The amount eventually allocated for drought relief was more than that set aside for irrigation under the SUDENE plan for that period, Cr$349 million.[57] Nine days after Medici's speech the Plan for National Integration was launched which officially announced the government's intentions for colonizing the Amazon region with emigrants from the drought-affected North-East. Led by the governor of Paraíba, João Agripino, the other North-Eastern governors expressed their concern about what they saw as the implicit neglect of long-term development strategy for their region which seemed to accompany these proposals. At least one contemporary observer saw the Amazon strategy as 'an escape mechanism by which the political system can no longer be captured by the close-knit North-Eastern bloc of politicians, once political activities such as direct election of governors are restored'.[58]

However, the latest government plans seem to restore the balance between the two regions. Under the Second National Development Plan (II PND) of 1974 resources are divided between the *Programa de Polos Agropecuários e Agrominerais da Amazônia* (*POLAMAZONIA*) and the *Programa de Desenvolvimento de Áreas Integradas do Nordeste* (*POLO-NORDESTE*). The latter concentrates specifically on the North-East and embraces rural development proposals for the humid valleys, the high *serras*, the *caatinga* or dry *sertão* and the coastal table-lands. Irrigation in the *sertão* receives considerable emphasis as 'playing a very significant role in the rural development strategy for the region'.[59]

Undoubtedly influenced by the human catastrophe of the 1970 drought and the inadequacy of measures to date in having long-term or wide-ranging impact, 1971 saw the publication of the first systematic irrigation plan for the North-East and other parts of Brazil, the thirteen volume *Programa Plurianual de Irrigação* (*PPI*). It was financed by funds from the First National Development Plan, and its proposals envisage irrigating 195,000 hectares of land in the North-East from 1971 to 1980, requiring a total investment of Cr$3,188.7 billion at 1971 prices. DNOCS, with responsibility for 78,000 hectares, has thirty-six projects under the plan. CODEVASF is responsible for the valley of the river São Francisco and will irrigate 112,000 hectares in seventeen schemes.[60]

The situation over goals is a little confusing, however, since in its own publications DNOCS states that it plans to irrigate 100,000 hectares by 1980, installing 24,000 rural families.[61] The *PPI* makes optimistic forecasts about the social and economic profitability of irrigation in the North-East, anticipating the creation of 350,000 jobs and the raising of agricultural production and incomes by substantial amounts.[62]

However, although irrigation was never expected to be a panacea for the problems of the 'drought polygon', results so far can only be classed as disappointing. By 1976 CODEVASF had only managed to start work on two projects and DNOCS on twelve. DNOCS had irrigated around 9,000 hectares or 9% of its 1980 target, and absorbed some 900 families, or 4% of the number it hopes to accommodate by the end of the decade.[63]

Notwithstanding the limitations imposed on any analysis of the DNOCS programme by the sheer slowness in getting schemes off the ground, it is my intention to clarify the types of irrigation strategy currently being pursued and to try and relate these to the specific problems associated with the drought. Before going on to analyse the characteristics of irrigation projects I wish, therefore, to examine in detail the exact nature of the 'drought problem', the *problema da sêca* which has so often been freely used to explain the backwardness of the North-East, but which has been subject to almost no systematic analysis. In chapter 2 I shall shift the emphasis away from the preoccupation with climatic uncertainties and more towards an analysis of the underlying social and economic weaknesses of rural society in the *sertão*, which exposes so many of its members to even minor rainfall variations. Observers of the 'drought problem' have traditionally concentrated on the climatic aspect, while ignoring the more fundamental problems of the way in which rural society is organized to resist the drought.[64] The perpetuation of what, I suggest, is largely a misconception, has been encouraged by North-Eastern writers such as Graciliano Ramos and Rachel de Queiroz, whose novels tend to focus on the punitive climate rather than the social roots of the drought phenomenon.[65]

I shall argue, however, that the so-called 'drought problem' is not only, nor even principally, a climatic problem as its name suggests. The human tragedy of the drought is a direct result of the way in which the rural structure of the *sertão* places thousands of peasants at the economic margin, vulnerable to even the slightest climatic vicissitude. After firstly examining the climatic aspect of the drought, and secondly identifying the vulnerable groups and the reasons why they are so adversely affected by dry periods, I shall proceed to an analysis of whether the 'drought problem', redefined in these terms, is being effectively tackled by current DNOCS irrigation strategy.

2. O problema da sêca

I. The climatic problem

There are two related but conceptually distinct facets to the 'drought problem' in North-East Brazil. There is the climatic problem which causes rainfall shortages or delays the arrival of winter rains essential for the growing of subsistence food crops. The 'drought problem' is not simply one of climate, however. What has to be explained is why the rural population in the *sertão* is so vulnerable to variations in the rainfall cycle. The other side of the 'drought problem', then, is not directly concerned with climate at all but is social and economic in nature. One of the main reasons for suggesting that the rural population of the interior is particularly vulnerable to the climate is that phenomena usually associated with full drought situations (such as that in 1970) are clearly observable even when the winter rains in the *sertão* are delayed by as little as a few weeks. In 1975, for example, the rains were delayed in many parts of the interior, crops failed, relief supplies had to be distributed and impoverished rural workers started migrating to seek employment in the towns. The arrival of the rains in time for sowing saw an immediate return to normality. The distinction between a drought situation and a normal situation is, therefore, a very fine one.

Section II of this chapter will examine in detail the social and economic characteristics of the *sertão* rural population, showing why some groups are protected from the drought effects, but how the majority of low-income producers, and some occupational groups in particular, are especially liable to suffer as the result of climatic vicissitudes. This section will be devoted towards clarifying the climatic aspect of the 'drought problem', an understanding of which is crucial to a broader comprehension of the drought's impact upon the rural population.

The *sertão* of North-East Brazil, often referred to in translation as 'the semi-arid backlands' or its equivalent, is far from being the barren desert that is sometimes suggested. It is, rather, a vast region of extremely diverse geography with considerable variations in topography and climate. Euclides de Cunha, in his classic work on the region, aptly named his book *Os Sertões* rather than *O Sertão*. Covering some 650,000 square kilometres, the *sertão* includes six major river systems as well as several fertile upland *serras*.[1] That part of the *sertão* usually referred to as 'semi-arid' is known as the *caatinga*, an indigenous word meaning 'white forest'. This covers the drier, less fertile lowland areas away from the floors of the river valleys. The *caatinga*, which starts on the valley-sides and extends over the surrounding hills, is characterized by drought-resisting xerophytic vegetation such as cacti, shrubs and small trees, The roots of trees such as the *umbuzeiro*, *jurema* and *juazeiro* accumulate water for the dry season which extends from June to December; the *caatingueira* loses its leaves during the summer months in order to minimize loss of water through evaporation. Numerous types of cacti abound which shed their leaves for the same reason.[2]

The mean annual rainfall for the *sertão* as a whole is quite high at 700 mm, but varies from 250 mm in the driest areas to well over 1,000 mm in the humid *serras*; as much as 1,700 mm, for example, has been recorded in the Serra do Baturité in Ceará.[3] This rainfall pattern permits extensive cattle-raising on the *caatinga* during the wet season (January to May approximately) until the dry months arrive, when the animals are usually transferred to the margins of rivers or reservoirs, or to the high *serras* for pasturing. In the river valleys a type of agriculture has grown up peculiar to the North-Eastern interior known as *agricultura de vazantes*. Under this system small farmers grow their food crops such as beans, rice, maize and some vegetables on the dried-up river beds or around the edges of reservoirs, enabling a second harvest to be made during the dry season when the nearby lands of the valley, whose fertility depends solely on rainfall, are unproductive. Drought-resisting plants form the basis of extractive industries which are very important in the rural economy of the *sertão*. They not only include *mocó* cotton, but also the wax-yielding *carnaúba* palm, the *oiticica* tree whose nuts yield oil and the *palma forrageira*, a spineless cactus which is shredded for use as cattle fodder. Up in the cooler, more humid *serras* a vast array of crops is produced: fruits, vegetables, manioc, sugar-cane, and even tobacco and coffee.[4]

The mean annual rainfall of the *sertão* compares favourably with many parts of Europe such as Paris, which has 650 mm, and London with 800 mm. The mean annual rainfall, however, has little significance in a region where distribution over the year is very asymmetrical and the median monthly rainfall is much higher than the average monthly precipitation. The climatic problem of this region is not, therefore, one of scarcity of rainfall in absolute terms but that of irregular distribution over the year. Even during a 'normal' year 80–90% of the rainfall is concentrated during the wet season. The duration of the rainy season is fairly constant but its starting point, which coincides roughly with sowing time in the agricultural calendar, may vary by between fifty-five and eighty-five days, with an 80% probability of correspondence. The reduction of total rainfall by one-third can have disastrous effects if the start of the rainy season is delayed for long enough to make the crops fail.[5] The distribution of rainfall can be so erratic that both drought and floods (*enchentes*) may occur in the same region within the space of two or three years. The effects of poor distribution are further aggravated by the high rates of run-off and evaporation of rainwater. In the *caatinga* especially, impermeable, crystalline rock formations are common, which slope towards the rivers, facilitating rapid run-off, soil erosion, silting up of rivers and evaporation.

The drought of 1970 presents a good example of the effects of these climatic features. It is sometimes assumed that a drought necessarily implies a severe decline in the total volume of rainfall falling within the affected area. It has even been suggested that the drop in precipitation is usually as high as 70% of the normal level during a drought year in the North-East.[6] However, evidence from the 1970 experience, typical in all respects of other serious droughts in the *sertão*, does not support this idea. A systematic analysis of the climatic changes which occurred during the drought found that the total volume of rainfall during the year had fallen substantially in only a few areas of the *sertão*, in Ceará and western Pernambuco. The losses in agricultural production that were suffered in 1970, it concluded, were due far more to the irregular distribution of rainfall during the rainy season than to a severe decline in the overall volume of rainfall.[7] The remainder of the *sertão* registered deviations of less than 25% from the normal mean annual rainfall, while some areas even showed a substantial increase (north-eastern Rio Grande do Norte, western Paraíba and parts of Bahia). Such a deviation of less than 25% is not significant in determining the intensity of a drought since this sort of decrease may occur once every two to five years; it is, in fact, representative of 'normal' climatic conditions in those areas. Once the deviation reaches 30–50% or more it becomes exceptional, occurring once every ten to twenty years, and is more likely to produce a drought situation.

However, this sort of absolute decline in total rainfall only affected limited areas of the *sertão* in 1970. A significant factor in the 1970 drought was the severe irregularity of rainfall between December 1969 and May 1970, during which time four relatively dry months alternated with two quite wet months to frustrate the agricultural calendar of subsistence farmers. In some areas of the *sertão* there was insufficient rainfall at the time of sowing, while in other parts the first rains were not followed up quickly enough by second and third rains, so that the crops could not mature in the ground. Harvests were lost and, once peoples' relatively small stocks of food were exhausted, they began to suffer the effects of the *sêca*. Landowners with cattle, on the other hand, felt the drought effects far less severely. Evidence from the above-quoted study of the 1970 drought clearly shows that medium and large landowners were far less adversely affected by the *sêca* than the mass of subsistence farmers and rural workers. This is explained in greater detail in section II of this chapter.

The relatively recent expansion of the *sertão* population which resulted initially from the nineteenth-century cotton boom has been described in chapter 1. The population of the region has grown steadily ever since, so that today the *sertão* rural population of eighteen million accounts for 60% of the total North-East population of just over thirty million. The rural population of the *sertão* tends to be concentrated in the humid valleys, away from the drier areas of the *caatinga*. In the valleys population density is quite high at around sixteen to seventeen people per square kilometre, and can be as high as twenty-five per square kilometre in more fertile areas such as the Carirí in Ceará. In drier areas such as the *sertão* of Crateus in Ceará, however, the figure drops to around ten people per square kilometre.[8] Nevertheless, taken as a whole, the population at risk during a drought is fairly substantial for a region which is often dismissed as infertile, semi-arid scrubland. These two factors, that is high population density and the effect of an erratic rainfall cycle on a precarious subsistence agriculture, go much of the way toward explaining the large-scale human impact of the drought in the *sertão*.

The particular vulnerability of the *sertão* rural population to irregular rainfall distribution was well illustrated during the first half of 1975 when the winter rains, which usually start in January–March, depending on the region, did not in many cases arrive until late April. In the *sertão* tradition holds that if it does not rain by 19 March, St Joseph's Day, a *sêca* is bound to follow. By the middle of April it was reported that many small farmers were in danger of losing a large part of their crops such as beans, manioc and maize, upon which they depend for their food supply.[9] During the same month North-Eastern state governors took part in a special meeting at SUDENE headquarters to discuss contingency plans for the setting up of work fronts if the situation deteriorated further.[10] In the state of Rio Grande do Norte it was reported that hungry peasants had tried to ransack food stores in a region where subsistence food stocks had reached a particularly low ebb.[11] However, during the last week of April the winter rains arrived and a drought situation was averted.

Again, in 1976, delay in the arrival of the winter rains in certain parts of the interior was enough to necessitate the distribution of emergency relief to the hungry rural population. As early as January it was reported that in the Irecé region of Bahia three sowings of beans and maize had failed and rations were being given out, although at that stage it was certainly not a drought.[12] In Ouricurí, situated in the *sertão* of Pernambuco, bus and trainloads of refugees made their way to São Paulo in search of employment. Many of the unemployed that remained were given jobs on work fronts.[13] For those with no resources to tide them over until the arrival of the rainy season the effect on their lives was similar to that imposed by a drought even though, technically speaking, it did not develop into a drought situation proper until later in the year. The events of 1975 and 1976 therefore demonstrate very clearly how delicate is the dividing line between normality and a drought for the rural inhabitants of the *sertão*.[14]

The *sertão* of North-East Brazil is, as we have seen, a region of highly diverse geography which has both drier areas of semi-scrubland and more humid, fertile valleys and highlands. References to its aridity and desert-like qualities abound in the writings of DNOCS engineers who have traditionally been more concerned with problems of water supply as a solution to the drought than with those of socio-economic factors such as landownership, labour relationships and access to rural credit. The preoccupation with climate has resulted in the more politically controversial questions of rural structure in the *sertão* being virtually ignored in official plans such as the *política hidráulica* of DNOCS. Climate is only one side of the problem and, as I hope to demonstrate in section II, the way in which the rural economy and society in the *sertão* are organized is as important, if not more so, in explaining the vulnerability of the *sertanejo* to the drought.

Nevertheless, the preoccupation of many observers with climate persists. The *sertão* was recently compared, for example, with the Sahelian–Sudanic region of Africa.[15] It is worth drawing out in some detail the fundamental differences between these perhaps superficially similar regions. This will illustrate the ambiguous nature of the Brazilian drought and its origins in socio-economic factors rather than climatic hardships. Any attempt to compare the *sertão* with true deserts should therefore be treated with some caution.

The Sahelian–Sudanic region of Africa lies south of the Sahara and extends through the former French colonies of Senegal, Mauritania, Mali, Niger, the Upper Volta and Chad, an area of two million square miles with a population of sixty million. During 1973–5 it was estimated that between five and ten million people were facing starvation as the result of drought in the western part of the region, generally known as the Sahel, an area with a population of about twenty-five million. In Ethiopia it is thought that between 50 and 100 thousand people died as a result of the drought. Comparisons between the Brazilian *sertão* and the Sahel are, however, very misleading, because drought in the African region is relatively abnormal.

The nomadic pastoralists from the northern Sahel, where rainfall averages a low 100–350 mm per annum, have over the centuries evolved a balanced system of exchange with the sedentary population of subsistence farmers in the south, who enjoy a mean annual rainfall of 350–600 mm per annum. During the dry season in the north the tribesmen move their herds of cattle to graze in the south and, in exchange for pasturing and watering rights, the animals deposit manure to fertilize the farmers' lands. Some migration southwards after the rainy season is therefore quite normal.

Between 1968 and 1973, however, a severe and progressive decline in mean annual rainfall took place over the whole Sahel. Livestock perished (50% in Mauritania, 40% in Mali), crops failed and pastures did not replenish themselves. This caused heavily inflated migratory movements of northern tribesmen southwards, increasing the already substantial pressure on the diminished resources there and creating conflicts over land and water rights. The important point as far as this study is concerned, however, is that this drought was abnormal, brought about by a progressive deterioration in climatic conditions over recent years due to external factors. Until 1968 there existed a balance between northern and southern agricultural populations which had enabled that in the former region to adapt itself very satisfactorily to the semi-arid climate.[16]

In the Brazilian *sertão*, on the other hand, the instability generally associated with droughts is quite normal, and is commonplace after even slight rainfall shortages. This is enough to produce catastrophic results in human terms precisely because the agricultural population has not adapted itself to the semi-arid climate and a large part of the low-income rural mass lives a highly precarious and unpredictable existence, totally dependent on their few subsistence crops to support themselves over the dry season. The balance between survival and starvation in the *sertão* is so delicate that even during a normal dry season many small farmers experience considerable difficulty in making ends meet.

In summary, the *sertão* population is vulnerable to even slight variations in the normal rainfall distribution whereas in the Sahel it took a substantial decrease in mean annual rainfall over a number of years to produce a drought situation. A series of external factors was responsible for this climatic deterioration: two theories advanced suggest that changes in the position of high pressure zones occurred recently and that there has been a change in the world distribution of temperatures, which caused the steady decline in mean annual rainfall in the Sahel. Pressures on the ecology have been further increased by other factors: pastures have been overgrazed because of a large increase recently in the number of cattle caused by, amongst other factors, cattle-health improvement schemes. Improper land-use characterized by slash-and-burn agriculture and deforestation has also left soil vulnerable to erosion while decreasing its fertility. Consequently, the water-table has been lowered by up to 50 or 100 feet (15–30 metres) in some parts of the Sahel, ruining oases and immobilizing hydro-electric installations.[17] As a result of these factors a process of 'desertification' is taking place in which the Sahara Desert is reputed to be advancing by as much as 30 miles (50 kilometres) a year in some areas, invading the semi-fertile zone. In the last fifty years it is estimated that the Sahara has claimed 250,000 square miles (400,000 square kilometres) of arable land.[18]

Generally speaking it may be said that these characteristics do not apply to the *sertão* of North-East Brazil. The mean annual rainfall is fairly consistent, with deviations of more than 25% an exception. The drought problem is, rather, one of a precarious subsistence agriculture dependent on rainfall which is subject to an irregular distribution throughout the growing season. Some observers have suggested, however, that the *sertão* has been subject to at least some of the features of 'desertification'. Overgrazing, slash-and-burn agriculture and poor soil-use are blamed for the disappearance of much of the vegetation that the *caatinga* used to support. This has facilitated soil erosion and has affected the hydraulic regimes of major rivers, making them more erratic. It is alleged, for example, that this is one of the reasons for the drying up of the tributaries of the river São Francisco and of its sharply decreased volume of water in 1970. Deforestation of large areas along its banks in order to provide fuel for paddle steamers over the past century has also, it is said, led to increased soil erosion which has made the river shallower. CODEVASF has, in fact, started a reafforestation programme in response to this phenomenon.[19]

There is an inherent danger, however, that if too much generalization takes place on the basis of such evidence, attention will be drawn away from the central problem, which has to do with neither the climate nor the ecology of the *sertão*. Sensationalist newspaper articles have already talked of the 'desertification of the North-East', with very little evidence to substantiate this claim.[20] Such arguments reinforce the popular image of the *sertão* as a region whose misfortunes are entirely the product of a harsh climate and poor natural resources, and whose salvation lies in a climatic solution. Thus much has been made of claims that it will soon be possible to create artificial rainfall in the interior, to the extent of being able to do away with the drought completely. One scientist recently proclaimed that artificial rainfall will 'resolve once and for all the drought problem in the North-East, modifying the climate of the region, reestablishing a perfect hydrological cycle and regulating rainfall'. This would be achieved by burning petroleum at strategic points in order to create carbon particles which would rise and form rain-bearing clouds that would be carried across the *sertão* by the wind.[21] The announcement of this technique inspired a major North-Eastern newspaper to proclaim 'The End of Droughts in the North-East'.[22]

Despite the enormous technical difficulties involved, the same assumption is present that control of water availability on its own is sufficient to solve the drought problem. This false assumption was, to a large extent, responsible for the failure of DNOCS's *política hidráulica*, based on reservoir construction, to have any substantial impact on the drought. DNOCS still places considerable emphasis on rainfall distribution as the main problem facing the *sertão*, comparing drought in the North-East to the 'world problem' of arid and semi-arid lands.[23] I wish, in section II of this chapter, to change the climatic emphasis somewhat, by analysing the rural structure of the *sertão* in order to show how its evolution and present-day characteristics influence the vulnerability of its population to climatic uncertainties.

II. The socio-economic problem of the drought

In this section I intend to demonstrate that the human tragedy of the drought in North-East Brazil is due as much to the social and economic organization of the region as to climatic vicissitudes. I shall argue that the agricultural economy of the *sertão* places certain clearly definable groups of rural producers in a situation of weakness and dependency which leaves them with few resources to fall back on in times of hardship when a drought occurs. The task of identifying these groups and accounting for their particular vulnerability is clearly central to any analysis of the problem. Any government strategy aimed at alleviating the impact of the drought, such as irrigation, must somehow reach these people if anti-drought measures are to have any substantial or visible counter-effect.

Some eight million people or 32% of the total population of the North-East falls within the boundaries of the drought-affected zone.[24] During a drought overall agricultural production declines significantly over previous years (see table 1), a loss which, in 1970, amounted to an estimated US$300 million, equivalent to a 17% decrease in production over 1969. In 1951 and 1958 the drop was in the order of 10%.[25] The drought creates such havoc amongst certain groups of the low-income rural population, however, because it is precisely those crops upon which they depend for

TABLE 1. *Variations in total production (factor cost), North-East Brazil, 1948–71*

Year	*Agriculture*	*Industry*	*Services*	*Total*
1948	4.7	4.2	2.5	3.1
1949	4.8	8.8	2.0	3.9
1950	4.6	3.3	5.4	4.8
1951*	–9.9	–5.9	8.2	–0.8
1952	4.8	8.4	11.0	8.7
1953	4.4	0.0	–5.3	–1.6
1954	12.4	10.0	14.4	12.9
1955	2.8	8.7	3.2	4.1
1956	5.3	21.0	4.9	7.4
1957	7.3	9.4	17.8	13.0
1958*	–9.7	10.0	2.5	0.3
1959	18.0	20.5	4.6	9.3
1960	7.6	1.4	14.8	9.9
1961	4.4	2.4	8.5	7.0
1962	7.6	–6.9	11.5	3.8
1963	6.6	2.7	4.6	6.6
1964	1.2	14.7	2.2	2.6
1965	8.4	1.9	3.4	4.9
1966	–0.9	–7.3	2.8	1.5
1967	14.8	4.2	8.9	10.2
1968	6.2	16.3	8.7	9.4
1969	1.3	8.4	10.4	7.3
1970*	–16.9	11.9	3.2	–0.9
1971	18.4	8.4	5.7	10.2

Source: Clovis Cavalcanti and Dirceu Pessoa, *Caráter e Efeitos da Sêca Nordestina de 1970*, SIRAC/SUDENE, Recife, 1973, p. 74. (Figures for 1948–67 are based on estimates by the Fundação Getulio Vargas; 1968–71 data is from SUDENE.)
*Indicates drought year.

their livelihood and their own diet that are most adversely affected, namely cotton and staple foods such as beans, corn and manioc. As table 2 shows, in drought-hit areas total production of these crops fell by over 80% in 1970. This fact is of crucial importance in explaining the severity of the social calamity since these staple foods make up the diet of large sections of the rural populace and represent their only means of support during a drought. This is especially true of sharecroppers (*parceiros*) and small-holders (*minifundistas*) who have remained, if not totally outside the money economy, connected to it only in an indirect fashion. That is, they are primarily subsistence farmers in the sense that most of their produce is consumed by the farmer and his family rather than entering the market.

TABLE 2. *Variations in the volume and prices of major crops during drought years*

Crops	*Volume produced (%)* 1970	1958	1951	*Prices (%)* 1970	1958	1951
1. *Crops vulnerable to the drought*						
Cotton	–87	–35	–44	+24	+45	+31
Corn	–86	–32	–27	+57	+11	+23
Beans	–78	–41	–18	+84	+17	+22
Manioc	–86	–11	–12	+113	+6	+14
2. *Crops unaffected by drought*						
Sugar-cane	NDA	–2	+3	NDA	+3	+2
Cocoa	NDA	0	–22	NDA	+16	+4
TOTAL	NDA	–16	–18	NDA	+18	+12

Sources: 1958, 1951. David Goodman and Roberto Cavalcanti, *Incentivos à Industrialização e Desenvolvimento do Nordeste*, IPEA/INPES, R.J., 1974, p. 117.
1970. Cavalcanti and Pessoa, *op. cit.* p. 175.
Note: Figures refer to percentage changes in volume and prices during drought years over preceding year.

This argument is developed in greater detail below, but it is worth making the basic point here that because of the small cash income received by these groups, either from the sale of their produce or from wage payments, they are primarily dependent on a few meagre stocks of staple foods to see them through long dry spells.

During a drought the scarcity of food supplies causes prices to rise considerably (see table 2). This has a compensatory effect for those producers (usually large and medium-sized landowners) and merchants who are able to store produce and can take advantage of their more direct links with the market economy to exploit the price increases. Many in fact make large profits from the shortages, as in the infamous *indústrias da sêca* already discussed in chapter 1.[26] Pressures on already scarce foodstuffs are increased by the entry of previously self-sufficient producers onto the market with what little money they have at their disposal. Work fronts relieve this pressure somewhat by injecting money into the drought-hit economy, but it is possible that they encourage landowners to dispense with their workers temporarily. This removes from employers the onus of having to support their dependents while minimizing the risk of losing labour through migration from the area.

In summary, then, sharecroppers and to a lesser extent smallholders are likely to suffer most during a drought because they lack the resources to maintain themselves on their plots. An important factor which influences the farmer's decision whether to migrate or not is the availability not just of food but of water. Shortages of permanent, local supplies of drinking water for human consumption provide an additional pressure on families to leave a drought-hit region. This problem is compounded by the practical difficulties of distributing sufficient quantities of the liquid in lorries (*carros pipas*) to isolated rural communities. Neither are smaller *fazendeiros* totally immune. If natural pastures and smaller *açudes* dry up in extreme cases, cattle deaths may be caused by lack of water and undernourishment due to inadequate fodder from cacti and leaves.[27]

It was recognized long ago that in North-East Brazil rural labourers, as opposed to landowners, are particularly vulnerable to the *sêca*. At the beginning of the nineteenth century the travellers Spix and Martius noted with alarm how landowners were reluctant to allow their workers to plant food crops on land which could be put to other use: '. . . owners of large *fazendas* do not wish to cede the smallest area of their land (for food production) because they consider these large tracts indispensable for cattle raising'.[28] More recently a government report recognized that the harmful effects of the drought are 'concentrated on a group with no means of defence'.[29] However, only in 1970 was the differential impact of the *sêca* on various rural groups analysed in any detail. Nearly 3,000 interviews were conducted on work fronts and agricultural establishments in the drought zone.[30] The results show quite clearly that sharecroppers (*parceiros*) are the worst affected, accounting for 42% of work front members but only 7% of the total rural labour force in the *sertão* (see table 3). Smallholders are the next most adversely affected group, constituting 31% of front recruits. According to Cavalcanti and Pessoa the impact on *minifundistas* is proportional to their membership of the rural labour force (30.68%). This figure is derived from statistics for the whole North-East, however, rather than just the interior. Information from the 1970 Agricultural Census permits a breakdown to be made and it emerges that the real figure is 19.29% (see table 3). The effects of the drought are thus felt disproportionately by both *parceiros* and *minifundistas*.

The drought has a far greater adverse effect on smallholdings than larger properties. Contrary to the opinion sometimes voiced that larger *fazendeiros* are more prone to evicting workers, in 1970 it was found that 85% of work front members came from properties of under 10 hectares.[31]

TABLE 3. *Origins of work front recruits compared with general occupational structure, sertão, 1970*

	Occupation					
Proportions (%)	Land-owner	Wage-worker	Tenant farmer	Share-cropper	Squatter	Other
1. Work front recruits	30.68	8.43	5.02	42.18	2.64	11.05
2. General occupational structure	19.29	8.0	3.25	7.2	4.65	57.61

Sources: 1. Cavalcanti and Pessoa, *op. cit.* table 5.6, p. 111.
2. Calculated from IBGE, *Agricultural Census*, 1970.

In the *sertão* as a whole, *minifúndios* in this category account for 61% of the total number of holdings and only 4.6% of the total cultivated area, although clearly the density of population is much higher in this group (see table 4). Furthermore, despite the fact that sharecroppers obliged to leave small properties formed the largest section of work front recruits in 1970, *parceiros* tend to be concentrated on *latifúndios.* The Agricultural Census does not break down forms of rural employment by property size, but a recent survey suggests that in the *sertão* sharecroppers make up 1% of the labour force on properties of under 10 hectares, rising to 25% on holdings of over 500 hectares.[32]

TABLE 4. *Origins of work front recruits by property size compared with general distribution of landholdings, sertão, 1970*

	Property sizes (ha)		
Proportions (%)	0.1–10	10.1–100	100.1+
1. Origin of work front recruits	85.09	12.94	1.97
2. General distribution of landholdings, sertão, 1970	61.0	35.59	3.41
3. Area covered	4.61	26.33	69.06

Sources: 1. Cavalcanti and Pessoa, *op. cit.* p. 114.
2 and 3. Calculated from IBGE, *Agricultural Census*, 1970.

The drought also discriminates geographically as well as occupationally and by property size. The areas of highest population density in the *sertão*, in the states of Ceará, Paraíba and Rio Grande do Norte, are the worst affected by the drought in terms of the proportion of the population recruited onto work fronts (see table 5). In addition, it is precisely these three states which have the highest proportion of subordinate sharecroppers in their rural work force (see table 11 below). A combination, then, of high population density and popularity of sharecropping helps explain why certain areas require inordinately heavy government emergency relief during a drought. This is corroborated by table 6 which demonstrates that it is again these regions which sustain the heaviest losses in both food and industrial crops during the drought.

Larger properties are protected from the drought by concentrating their assets into cattle-rearing and cotton cultivation. Extensive cattle-raising formed the economic basis for the colonization of the *sertão* and it has always been relatively resistant to the rigours of periodic droughts. This is not to say, of course, that *fazendeiros* were such a privileged elite as to be totally immune. Cattle ranches were not, on the whole, very wealthy in absolute terms. Their herds were fairly small with a low annual production and the cattle left to graze on the open range for most of the year. Most *fazendas* in the São Francisco valley, for example, produced an average of fifty to one hundred calves per year in the 1860s, while the figure was nearer thirty per annum in Rio Grande do Norte. If we accept contemporary evidence, it seems that most of the profits from the sale of cattle went into the pockets of the middlemen or *marchantes* rather than the landowners'.[33]

Nevertheless only the most severe droughts such as that of 1877–9 seriously depleted the cattle population. Cattle has the advantage of being a geographically mobile asset so that, during a dry period, owners could move their herds to the more humid valleys and *serras*. These *retiradas* of animals were a common feature of nineteenth-century life in the interior.[34] Nevertheless, transport difficulties were greater then and the mobility of ranchers with fewer resources more limited. Until the late nineteenth century additional watering facilities were extremely scarce. Nowadays many cattle raisers have access to better grazing and watering for their animals near private or public *açudes*, as well as motorized transport for transferring their livestock to better pastures should the need arise. *Fazendeiros* also switch areas of cotton and staple food production to grazing in order to protect their cattle.[35] In this way the burden of the drought falls unequally on the shoulders of the sharecropper who, faced with the extra problem of trying to eke out an existence on drier soils, sees his means of

TABLE 5. *Geographical origins of work front recruits, sertão, 1970*

				No. work fronts		*No. flagelados*		
States	*% area affected by drought*	*% population affected*	*% municipality affected*	Total	% N.E.	Total	% N.E.	% state population
Piauí	46	37	44	8	6.8	28,772	7.3	1.66
Ceará	74	57	90	23	19.5	160,984	40.9	3.59
Rio Grande do Norte	92	54	85	15	12.7	70,450	17.9	4.37
Paraíba	74	45	61	16	13.6	75,657	19.2	3.10
Pernambuco	72	21	43	48	40.6	49,565	12.6	0.94
Alagoas	44	21	35	4	3.4	5,052	1.3	0.31
Sergipe	43	21	37	3	2.5	2,105	0.5	0.23
Bahia	24	16	18	1	0.8	1,002	0.3	0.01
SERTÃO	46	32	48	118	100.0	393,587	100.0	1.53

Source: Cavalcanti and Pessoa, *op. cit.* p. 77.

TABLE 6. *Differences in the value of agricultural production at constant prices in drought years compared with preceding year (%)*

	Food crops			*Industrial crops*		
States	1951	1958	1970*	1951	1958	1970*
Piauí	−24.8	−18.0	−21.4	−17.5	−35.8	−26.6
Ceará	−38.8	−65.4	−52.1	−67.3	−62.5	−64.9
Rio Grande do Norte	−10.8	−27.6	−19.2	−34.3	−58.3	−46.3
Paraíba	−20.1	−28.1	−24.1	−28.2	−16.3	−22.2
Pernambuco	−12.2	−11.3	−11.7	−1.0	−12.4	−6.7
Alagoas	−0.2	−3.4	−1.6	−8.8	−28.3	−18.5
Sergipe	−12.3	−6.8	−9.5	−22.8	−1.3	−12.0
Bahia	−10.8	−8.0	−9.4	−3.3	−1.6	−2.4

Source: José Otamar de Carvalho, *Plano Integrado para o Combate Preventivo Aos Efeitos das Sêcas no Nordeste*, MINTER, Brasilia, 1973, p. 158.
*1970 figures are an average of 1951 and 1958.

support being consumed by foraging animals. As a subordinate worker with no legal or other power to stop this process, his only alternative is usually to leave the land.

The degree to which agricultural activity in the *sertão* is organized to protect cattle rather than other sectors during a drought is illustrated by the fact that while staple food production declined substantially in 1970 (table 2), cattle showed a slight increase over the previous year, in line with the national trend (table 7). The price of cattle went down slightly in most states but this was a temporary fall in their value which was not prolonged. The main impact of the drought on cattle producers seems to have taken the form of a short-term disincentive to invest further in cattle stocks and permanent crops.[36] The protection afforded to cattle producers during a drought is well illustrated by table 8, which shows that during 1970 large estates were able to switch their resources rapidly into livestock, as the situation demanded. The income from cattle of properties under 10 hectares decreased from 28% to 24%, while the income from livestock for all establishments rose from 27% to 46% of the total.

Larger properties also have the advantage of having a substantial portion of their commercial assets in cotton-growing, which complements cattle-raising rather well. Cotton is harvested in the dry season, leaving the land free for grazing cattle when the natural pastures dry up. Cotton may be

TABLE 7. *Size and value of cattle herds (bovine) 1969 and 1970*

	Head (000s)			*Value (cruzeiros)*	
States	1969	1970	1970/1969 %	1969	1970
Piauí	1,792	1,755	98	128	244
Ceará	2,251	2,227	99	184	343
Rio Grande do Norte	819	896	109	273	261
Paraíba	1,414	1,418	100	256	228
Pernambuco	1,539	1,534	100	264	246
Alagoas	810	819	101	310	274
Sergipe	824	836	101	283	272
Bahia	8,317	8,560	103	225	213
North-East	20,141	20,379	101	215	242
BRAZIL	95,150	97,864	103	196	210

Source: Cavalcanti and Pessoa, *op. cit.* p. 93.

interplanted with short-cycle crops such as beans and corn, permitting the development of sharecropping arrangements. Finally, cotton seed cake and corn husks provide valuable cattle fodder. During a drought, however, cotton production declines for two reasons: firstly, the land is obviously less productive and, secondly, having land under cotton normally allows this area to be converted into pasture when returns are maximized by grazing cattle there rather than using it to produce a small cotton crop. This decline in cotton production is clearly shown in table 2, which also indicates that producers received compensation for reduced output in the form of price rises. *Latifundistas*, who deal directly with wholesalers or even with cotton gins, are less likely to suffer a drop in income than smallholders and sharecroppers who are often obliged to deal through intermediaries and are more likely to receive either a smaller price than is their due, or to be paid in kind rather than cash.

The 1970 study suggests that the real income of properties below 25 hectares suffered most during the drought. Holdings of between 25 and 1,000 hectares, on the other hand, seemed to be the least affected.[37] These conclusions can only be tentative, however, since methodological difficulties of limited sample-size and of quantifying 'subsistence' food production do not allow a complete comparison of income according to property size. Nonetheless, as a general rule it seems reasonable to assume that larger holdings are better protected from the *sêca* by concentrating

TABLE 8. *Income composition of production units by property size, sertão, 1969–70 (%)*

Property sizes (ha)	Temporary crops	Permanent crops	Cattle	Milk	Wages	Total
1969						
under 10	39.2	10.9	28.6	17.4	3.7	100.0
10 – 24.9	32.7	11.6	29.3	25.7	0.5	100.0
25 – 49.9	13.2	4.7	18.0	63.7	0.2	100.0
50 – 99.9	19.4	7.2	34.1	38.8	0.3	100.0
100 – 999.9	24.7	11.8	29.3	34.0	0.1	100.0
1,000–9,999.9	48.6	9.6	23.4	18.2	0.0	100.0
TOTAL	30.6	10.3	27.1	31.4	0.4	100.0
1970						
under 10	10.6	52.3	24.3	10.4	2.2	100.0
10 – 24.9	15.8	8.4	47.7	27.2	0.6	100.0
25 – 49.9	10.9	2.4	29.6	56.7	0.2	100.0
50 – 99.9	20.7	3.0	42.5	33.4	0.2	100.0
100 – 999.9	6.9	6.1	53.1	33.5	0.1	100.0
1,000–9,999.9	7.7	10.7	55.6	25.9	0.0	100.0
TOTAL	9.3	12.6	46.4	31.1	0.4	100.0

Source: Cavalcanti and Pessoa, *op. cit.* p. 197.
Notes: 'Income' refers to gross real income (*renda bruta real*) which corresponds to the total volume of agricultural production, as well as wage payments and other income received by the head of household.

their activities on selling cattle and its by-products. Cotton also provides an income as well as an alternative source of grazing land at the expense of staple crops. The SUDENE/IBRD survey also documents the higher net income of larger properties. Despite having to support a far denser population, holdings of less than 10 hectares enjoyed an income equal to 11% of those in the 50 to 100 hectare group and to a mere 2% of properties over 500 hectares.[38] An additional consideration in the case of *fazendeiros* is that many have alternative incomes from interests outside of agriculture. Their dependence on a drought-affected agriculture is thus likely to be less than for the mass of rural workers.

The vulnerability of particular groups to the drought is a direct product of the landownership structure in the *sertão* and the system of tenancy relationships which has grown up around it. The commercializable rural surplus (cattle, cotton and some staple foods) is extracted by a relatively small minority of large and medium-sized landowners as well as a variety of merchants or middlemen, while a large part of the poorer population remain dependent on a precarious, largely subsistence agriculture susceptible to minor reductions in rainfall.

Landownership in the interior is very heavily skewed. Table 9 breaks down rural establishments into size categories for the *sertão*, showing that 58% of holdings are under 10 hectares and occupy less than 5% of the total area. At the other end of the scale, properties of over 100 hectares account for only 8% of the total number but cover 69% of the area. Criticisms have been levelled at Brazilian Census statistics on the ground that they understate the degree of concentration of ownership.[39] Use of 'rural establishment', an administrative division, rather than 'rural property', a juridical unit, does not take into account multiple ownership of several holdings administered by third parties on the owners' behalf. The latter criterion is that adopted by the *Instituto Brasileiro de Colonização e Reforma Agrária* (INCRA) in their surveys of rural areas but, even so, both the Brazilian Census Division (IBGE) and INCRA deal with continuous areas of land under the same administration or ownership respectively. No survey has yet shown the distribution of total areas of land owned by the same person, whether the properties are together or separated.

However, comparison of Census statistics with data obtained from independent studies of selected areas in the *sertão* suggests that IBGE estimates reflect fairly closely the real distribution of total property ownership. In the valley of the river Piranhas, Paraíba, it was found that 72% of properties were under 10 hectares but occupied only 19% of the land, the average size being 3.6 hectares. Larger holdings of more than 50 hectares, only 4% of the total number, took up 47% of the area, with an average size of 160 hectares.[40] In the neighbouring state of Ceará in the valley of the river Banabuiú 67% of the land was owned by the largest 16% of property holders.[41] Along the Jaguaribe this pattern is repeated, where 30% of all establishments are under 10 hectares and occupy a mere 3% of the land while those of over 100 hectares account for 10% of the number but nearly 50% of the area.[42] Similar conditions are to be found in all the densely populated river valleys of the *sertão*.

The severely unequal distribution of property ownership in the interior is a product of the fact that, from the earliest days of colonial expansion, the cattle *fazenda* was the basic unit of occupation and territorial expansion in the interior. Until the nineteenth century the *sertão* economy was based exclusively on extensive cattle-rearing using open-range methods on relatively large properties. The most common form of dividing the land amongst *donatários* was for each to be granted a distance of river frontage.

The boundaries of the property would extend vertically backwards from the water, frequently for an unspecified distance. Each landowner would have a section of wetter, more fertile soils as well as a slice of *caatinga* where his cattle could be pastured in winter months. During this phase the *sertão* was sparsely populated. The typical *fazenda* would employ a *vaqueiro* and perhaps some assistants on larger ranches. The only other components of the rural labour force consisted of a few peasants who were allowed to squat on the land if they agreed to provide food for the ranch. Farming was merely an appendage of the cattle industry.

During the nineteenth century, however, the situation changed radically. Cotton became established as a commercially successful crop, encouraged by a 32-year period without droughts in the *sertão* between 1845 and 1877 and boom conditions on the world market resulting from shortages caused by the American Civil War. Food-crop farming for the subsistence needs of an increasingly large rural population thus came to constitute an important activity in its own right. Simultaneously with this vast influx, properties were becoming more fragmented owing to the absence of primogeniture. In 1867 Sir Richard Burton observed that three generations of the same family had divided up an estate which used to run along 11 leagues (66 kilometres) of the river São Francisco into many smaller farms.[43] Orville Derby reported the same phenomenon in the area around the Paulo Affonso falls caused by inheritance and land sales.[44] As a result of this fragmentation process many landholdings are nowadays too small to provide a livelihood for their owners. In the Jaguaribe valley, for example, properties exist of less than one metre in width which stretch back from the river for perhaps several hundred metres. The most extreme case encountered was a property only 20 centimetres wide![45]

This period of expansion in the interior saw the emergence of a variety of tenancy arrangements between landowners and rural workers. In addition to the *vaqueiros* and the few small farmers who provided food for the large estates new categories were formed: tenant farmers (*rendeiros*), sharecroppers (*parceiros*), day labourers (*jornaleiros*) and squatters (*ocupantes*). These categories were adopted from the neighbouring *agreste* region closer to the coast which had been settled far earlier. These relationships had the effect of extracting for the landowners that portion of production which could be sold for a cash profit, namely, cattle, cotton and some food products. At the same time they gave the peasant the right to occupy the land in return for payment either of a cash rent or, as was more usually the case, for a stipulated share of the crop. On the one hand agricultural production was clearly linked to capitalist markets and was stimulated by the profit motive as far as the *fazendeiro* was concerned. On the

TABLE 9. *Changes in the distribution of landholdings, sertão, 1940–70*

	Size of landholdings (ha)											
	0–1.9		2–9.9		10–99.9		100–999.9		1,000+		*Total*	
States/years	N	A	N	A	N	A	N	A	N	A	N	A
Pernambuco												
1940	8.44	0.23	38.26	4.80	43.34	31.66	8.93	46.73	1.03	16.58	100	100
1970	15.91	0.50	41.89	5.46	34.98	31.03	6.91	46.03	0.31	16.98	100	100
Paraíba												
1940	2.27	0.04	26.32	1.12	59.32	28.16	11.33	49.54	0.76	21.14	100	100
1970	9.90	0.30	42.62	5.46	40.32	31.16	6.70	29.91	0.46	33.17	100	100
Rio Grande do Norte												
1940	3.49	0.04	18.65	1.12	55.96	19.58	20.19	49.54	1.71	29.27	100	100
1970	16.39	0.26	34.06	3.03	39.50	21.24	9.24	43.43	0.27	32.04	100	100
Ceará												
1940	5.18	0.05	18.48	1.07	55.89	22.11	19.19	51.24	0.55	25.53	100	100
1970	9.24	0.19	36.71	3.41	40.78	27.16	9.62	44.23	3.65	25.01	100	100

Piauí												
1940	5.66	0.06	32.22	1.42	39.01	11.69	20.88	47.81	2.23	39.02	100	100
1970	44.25	0.98	25.69	2.41	21.91	18.18	7.08	43.20	1.07	34.16	100	100
Bahia												
1940	16.33	0.62	48.21	6.97	29.20	26.54	5.54	40.32	0.72	25.55	100	100
1970	22.00	0.95	42.79	7.98	30.79	36.13	4.22	40.29	0.20	14.65	100	100
Sergipe												
1940	11.48	0.18	33.74	2.79	43.86	22.87	10.57	38.89	0.35	35.09	100	100
1970	19.43	0.38	26.24	2.78	45.57	31.84	8.38	42.96	0.38	22.04	100	100
Alagoas												
1940	10.32	0.66	54.31	15.19	32.73	43.78	2.49	26.70	0.15	13.67	100	100
1970	39.24	2.14	34.89	9.30	22.78	36.70	2.91	38.97	0.18	12.89	100	100
TOTAL SERTÃO												
1940	6.85	0.10	29.82	2.20	41.70	22.60	13.88	47.94	7.75	27.16	100	100
1970	22.24	0.54	35.88	4.07	32.28	26.33	7.27	43.56	1.33	25.50	100	100

Source: IBGE, *Agricultural Census*, 1940 and 1970
Notes: N = % of rural establishments
A = % of area covered

other hand it was only possible to extract this surplus through the maintenance of what may be termed 'feudalist-type' relations between landowners and their work force such as debt-bondage, reinforced by ties of *compadrasco* or godparenthood and patron-clientage. The majority of rural labourers had few direct dealings with the cash economy, paying their rent and receiving their rewards predominantly in kind.

It has been argued that in the nineteenth-century *sertão* the wealth of the landowning class depended on the area of land owned and on maintaining the lowest standard of living for the rural mass in order to maximize profits. Monopolization of the land, it is alleged, made it possible to determine the conditions of the poor by controlling their access to it.[46] Whether or not this was a conscious strategy on the part of most landowners is debatable. Certainly during this period the identifiable rural groups which were formed seem to have led a very precarious existence. This is true of squatters, sharecroppers and day-labourers, the landless ambulatory *agregados* who constituted the largest and poorest section of the peasantry. Tenant farmers seem to have been better off.

The economic position of low-income farmers in the *sertão* was aggravated by increased pressure on land caused by population growth and migration. Gradually the most fertile areas in the valleys and uplands became occupied, a process which was accelerated by the 1845 drought. In 1850 the *serras*, which until then had been the territorial patrimony of the indians, became public domain. By 1860 small farmers had reached the stage of invading the more arid *caatinga*, while settlers too poor to buy land became tenant farmers, squatters, sharecroppers and labourers for the landowners, forming a pool of cheap labour. The last three groups, the migratory *agregados* with no fixed contracts on *fazendas*, were particularly affected by the drought given their less secure positions.[47] The disastrous impact of the 1877–9 *sêca* may partly be explained by the high proportion of *agregados* employed in agriculture, an estimated 200,000 out of a total population of half a million in the state of Ceará alone.[48]

An additional strain on the rural work-force was produced by changes in the composition of the peasantry resulting from the cotton boom. Evidence suggests that many herdsmen and small farmers producing for their own subsistence needs switched to growing cotton, becoming marginal commercial farmers. The end of the boom period coincided roughly with the end of the 1877 drought, leaving farmers with a severely reduced income from cotton as well as fewer staple food crops with which to sustain themselves.[49]

Apart from the economic and social mechanisms underlying the system of domination, there existed an important political dimension. *Coronel-*

ismo is the name given to the system of political behaviour which allowed individual landowners or cliques of *fazendeiros* to control the vote in rural areas in exchange for favours from local, state and national elites. 'Colonels' were originally imperially appointed officers of the now extinct *Guardia Nacional*, a post either given to or bought by the most powerful landowner in a *município*. After the military post had been abolished the term was used as an unofficial title for local bosses.[50] First established in the sugar plantation society of Pernambuco and Bahia, *coronelismo* spread with the expansion of the agricultural frontier into the *sertão*. The power of a local *coronel* depended on the number of votes he could direct, frequently reinforced by various extra-legal methods such as fraud, threats of violence and the use of hired henchmen (*capangas*) to intimidate victims. Not only was he immune from the police, he frequently used them for his own personal ends, administering justice as he saw fit.[51]

At a broader level, then, the local boss was able to use his influence to exert pressure on state legislatures, party machines and juridical procedures as well as to gain access to additional financial resources such as bank credit.[52] The example of the Inhamuns area in Ceará illustrates how effectively one family could dominate the economy and political life of a region for over two hundred years despite the penetration of external threats to its authority such as the arrival of the police corps and the creation of political parties.[53]

In times of extreme need the colonels were not averse to recruiting the services of independent bandits (*cangaçeiros*) to supplement their own strong-arm men. If really desperate they would ally with each other, a case in point being the 1911 *pacto dos coronéis* in the Carirí valley.[54] At the same time, however, class solidarity amongst *coronéis* was undermined by often profound vertical splits when families would fight each other for control of local political offices or land. Despite the system of hierarchical interdependence of local, state and national levels of patronage, however, ultimately the power of the *coronel* rested on the dependence of his rural followers and the amount of grass roots support he could muster.

The significance of *coronelismo* at the level of the estate was that political and socio-economic controls reinforced each other. It has been suggested that this political activity was an essential partner to economic exploitation, allowing landlords to squeeze the maximum possible surplus from their work-force, eliciting submissiveness and crushing any resistance or attempts to challenge their monopoly over the land.[55] While no doubt there is a good deal of truth in this argument, it can be carried too far. If a landowner's degree of political leverage did depend on the number of votes he controlled and hence on the number of followers, he would surely risk losing his dependents to another patron if he treated them too harshly. There must have been some freedom of movement, although this was probably reduced by the coercive power of the landowner, by debt-bondage, difficult communications and similarity of working conditions overall.

In turning to an analysis of the existing system of tenure arrangements, principally those of sharecropping and *minifundismo*, it is precisely the more subtle forms of social and economic coercion, along with the more obvious pressures, which determine the dependence of these groups and their consequent vulnerability to the drought. Firstly, however, a few comments are in order on the problem of actually classifying rural occupational groupings. It is often rather difficult to establish hard and fast definitions between the various groups such as smallholders (*minifundistas*), tenant farmers (*arrendatários*), sharecroppers (*parceiros*), squatters (*ocupantes*) or wage-labourers (*assalariados* or *diaristas*). In theory a tenant farmer pays rent in cash while a sharecropper gives a certain portion of his crops to the landlord. In practice, however, both groups sometimes pay a combination of goods, cash or services. Similarly, a smallholder who cannot support his family from his plot of land may supplement his income by working as a day-labourer for a local *fazendeiro*. In one survey it was found that 4 out of 608 *parceiros* interviewed paid money as rent in addition to supplying goods and services to the landlord. Of seventy *arrendatários* forty-four paid exclusively in cash while the remainder supplemented this with labouring and part of the crop.[56] Another study of the *sertão* of Canindé, Ceará, discovered that 36% of sharecroppers supplemented their income by working as wage-labourers at some stage in the year.[57]

Despite these apparent anomalies, it is felt that distinctions between rural groups are drawn sufficiently clearly in the major sources of reference used in the present study. Particularly relevant is the *parceiro*'s relationship with the landlord for, as the 1970 drought survey by Cavalcanti and Pessoa showed, it is this group which is especially badly hit by the drought. All sources quoted adopt the standard definition of *parceiro* as a worker who receives payment mainly in kind and who pays rent in the form of an agreed share of the various crops (one-half, one-quarter, one-third, etc.). This is also the criterion adopted by IBGE and allows comparison of micro-studies with broader trends as indicated by official statistics.

An important conceptual problem which cannot be overlooked, of course, is how to establish a sharecropper's exact degree of dependence on a particular landowner. Some indication may be given by breaking down payments of goods, services and cash to the landowner. Alternatively, it may be possible to show, as does the Agricultural Census, whether the *parceiro* is 'responsible' for administering the property or whether he is subordinate to someone else. Ideally, a number of variables should be taken into account which have a bearing on the sharecropper's situation but which are difficult if not impossible to quantify using standard questionnaire survey techniques. These may include pressures such as intimidation or debt-bondage. Descriptive evidence has, therefore, to be used to complement quantitative data. The fact of *parceiros*' economic weakness and disproportionate vulnerability to the drought is, nevertheless, well established.

Sharecropping is more common in the *sertão* than in any other region of the North-East. A study of Pernambuco showed that, as a percentage of total cultivated area, twice as much land is sharecropped in the interior as in the more fertile *agreste* (see table 10). In terms of the percentage of rural establishments administered by sharecroppers as opposed to *arrendatários*, *parceria* in the Sertão do Pajeú is 2.2 times as common as in the *agreste setentrional*. For the interior as a whole, one and a half times as much land is sharecropped as is rented by cash-paying farmers.[58] In the coastal *zona da mata* sharecropping is virtually non-existent.

About 83% of agricultural labour in the interior is provided by the person administering a property and his family (see table 11). The landowner who wishes to recruit additional workers on a longer-term basis than temporary *assalariados* employed during busy periods such as the harvest, is faced with few alternatives. The other major tenancy arrangement apart from sharecropping would be to lease the land to a tenant farmer or *arrendatário*. In 1970, according to the Agricultural Census, of 500,000 'responsible' producers in the *sertão* some 84,000 consisted of tenant farmers, or 3.25% of the total labour force. The number of sharecroppers (both 'responsible' and subordinate) totalled 186,000 or 7.2% of the *sertão* rural labour force of 750,000. Although many landlords employ tenant farmers, then, *parceria* is clearly the most common tenure arrangement after ownership of land itself (19.3% of the work-force).[59]

The sheer force of tradition could be used as an argument to help explain the continued use of sharecropping in the *sertão*. Ideally one would like to undertake a comprehensive analysis of the economic rationality behind the use of *parceria* in this region of Brazil. However, such an analysis is beyond the scope of the present study. It seems reasonable to suggest, nevertheless, that *parceria* offers certain advantages over other types of arrangement. Broadly speaking, under this system the landlord is able to extract a substantial economic surplus in kind which he is then able to market. This effectively minimizes the sharecropper's participation in the market economy and helps to stabilize employees on the land. The

TABLE 10. *Rural areas by types of tenancy arrangement: the sertão and agreste of Pernambuco, 1970 (%)*

	Own land				*Land cultivated by others*				*Total*			
	Type of cultivation				Type of cultivation				Type of cultivation			
Region	Direct	Share cropped	Rented	*Sub-total*	Direct	Share cropped	Rented	*Sub-total*	Direct	Share cropped	Rented	*Total*
Sertão do Pajeú	90.6	8.1	0.9	99.6	0.3	0.1	0.0	0.4	90.9	8.2	0.9	100.0
Agreste setentrional	84.8	3.9	10.7	99.4	0.5	0.1	0.0	0.6	85.3	4.0	10.7	100.0

Source: SIRAC, *Estudo da Rentabilidade dos Estabelecimentos Agropecuários e das Condições do Arrendamento da Terra no Estado de Pernambuco,* SUDENE/DAA, Recife, 1974, tables 3.6 and 3.9.

owner's share of the crop may vary considerably, but a typical example is the case of *parceiros* in the valley of the river Piranhas, Paraíba, who occupied 78% of the land rented out by landowners. The latter were generally paid one-half of the permanent crops (cotton, bananas) as well as one-third to one-fifth of all temporary crops (beans, corn, manioc), the remainder being for the sharecropper and his family to consume, sell or exchange.[60]

Johnson concludes that sharecropping is an adaptive form of production which enables *parceiros* to meet subsistence needs and make the best of things in their struggle for existence.[61] Cheung also points out in relation to Taiwan that sharecropping is not necessarily an inefficient way of allocating resources but may be a rational means of maximizing wealth under given circumstances.[62] Johnson argues, furthermore, that a *parceiro* can express his independence from a landlord by being able to leave the *fazenda* quite freely if he wishes, confident that he can find work with another landowner.[63] However, freedom of movement does not necessarily signify freedom from the system of *parceria*, which studies indicate to be similar in its basic features all over the *sertão*. Even Johnson admits that there are, in practice, considerable restraints on the sharecropper leaving the *fazenda* of his own free will. He has invested his labour in planting cotton and other crops which have a value and can be sold to the landlord or consumed. If the worker suddenly leaves he forfeits the right to any compensation, a restriction which was felt by most except the newly arrived on the farm that Johnson studied.[64]

A large degree of freedom of movement is only theoretically possible in areas where labour is scarce and its supply relatively inelastic, as appears to be the case in the area of Ceará studied by Johnson.[65] It is doubtful, on current evidence, whether this region is typical of the *sertão* as a whole. Firstly, several studies indicate a surplus of labour, especially in the more densely populated valleys.[66] Secondly, the degree of rural underemployment in the *sertão* is hidden by the fact that the unpaid, family labour sector absorbs much of the excess which remains after emigration. The importance of this group should not be underestimated since 1970 Census figures indicate that it accounts for 83% of the rural labour force in the *sertão* (see table 11). A recent study estimated that in 1970 the 'surplus' rural labour force was as high as 547,000 or 21% of the total.[67] The presence of such additional labour diminishes the bargaining power of labour generally because it acts as a reserve of manpower which can be tapped by landlords when required.

For the landowner, then, especially the larger *fazendeiros*, sharecropping is a reasonably profitable means of acquiring labour for his property. The

TABLE 11. *Changes in the rural labour structure, sertão, 1960–70 (%)*

States/years	Total	1. Unpaid family labour	2. Hired labour Permanent	Temporary	3. Sharecroppers (subordinate)	4. Other forms
Pernambuco						
1960	100.0	63.96	2.71	29.54	2.47	1.32
1970	100.0	88.24	1.55	9.13	0.67	0.41
Paraíba						
1960	100.0	62.55	4.36	18.00	12.28	2.81
1970	100.0	77.28	3.24	5.98	12.82	0.68
Rio Grande do Norte						
1960	100.0	51.32	4.77	20.96	20.46	2.49
1970	100.0	66.76	3.88	9.12	18.85	1.99
Ceará						
1960	100.0	49.72	4.33	33.58	8.29	4.08
1970	100.0	73.42	1.94	12.26	5.23	7.15
Piauí						
1960	100.0	67.85	2.49	24.14	1.96	3.56
1970	100.0	97.42	0.17	1.44	0.61	0.36
Bahia						
1960	100.0	78.90	3.32	15.67	0.90	1.21
1970	100.0	91.55	1.34	5.63	0.33	1.15
Sergipe						
1960	100.0	72.77	2.74	22.63	1.56	0.30
1970	100.0	94.56	1.05	4.02	0.14	0.23
Alagoas						
1960	100.0	69.81	4.13	24.39	0.56	1.11
1970	100.0	90.92	1.38	7.05	0.10	0.55
TOTAL SERTÃO						
1960	100.0	61.54	3.64	25.92	6.24	2.66
1970	100.0	82.78	1.75	7.99	4.58	2.90

Source: Calculated from IBGE, *Agricultural Census*, 1960 and 1970.

Notes:

1. Unpaid family labour – the person responsible for administering the property as well as members of his family who help without payment of any kind.

2. Hired labour – persons, including family members, who are paid either in cash and/or with part of the produce. This is divided into long term (permanent) and short term (temporary) arrangements.
3. Sharecroppers – Sharecroppers subordinate to the administration of the establishment who are paid with a share of the crops they produce. Non-subordinate share-tenants who are responsible for administering the establishment are included within the first group of unpaid family labour. This distinction is useful since subordinate sharecroppers are far more vulnerable to the drought than any other group.
4. Other forms – all other types of labour.

Similar criteria were used for both 1960 and 1970 Agricultural Censuses, which facilitated comparison. 1940 and 1950 figures adopt different definitions.

payments in kind mentioned above as the basic feature of *parceria* mean that the landlord may keep his workers on the land throughout the year without having to pay wages for the entire period, as he would have to do with *assalariados permanentes*. This system also reduces costs of production since the *parceiro* is inevitably required to pay for such items as seeds himself. In order to do this he will probably request a loan from his employer against the subsequent crop. At this stage the sharecropper has made a medium- to long-term obligation towards his landlord which places the former in a decidedly subordinate position. The overwhelming majority of agreements between sharecroppers and landlords are verbal, and liable to different interpretations according to who happens to have the upper hand.[68] Committed to repaying the loan, the *parceiro* is frequently forced to sell his share of the harvest to the landlord while it is still in the ground (*venda na folha*) at a price well below that which the employer will obtain on the market. If he sells his produce immediately after the harvest the tenant will still get a low price because supply is at its maximum and the *parceiro* is unable to store his cotton or beans for sale during the off-season.

Sharecropping is not a totally repressive relationship with no advantages for the *parceiro*. For the landless worker *parceria* represents a more attractive proposition than daily wage-labour which is unpredictable in supply and whose rewards are subject to inflationary pressures.[69] The sharecropper, unlike the *assalariado*, does not have to buy his major food supplies, and one of the advantages of receiving payment in kind is that it acts as a hedge against inflation.[70] Such benefits may be offset by weather risks or by competition amongst labour in reducing either land allotments or crop shares, but such divisions are usually fixed by tradition rather than by competitive forces.

On the whole sharecroppers in the *sertão* are better off than wage-labourers but worse off than tenant farmers or smallholders in terms of total income. A recent analysis of *parceria* in the North-East concluded that they fared better than *assalariados* despite the instability of employment and disincentives to modernization inherent in sharecropping.[71] In the *sertão* of Canindé, Ceará, the net family income of *parceiros* was found to be 28% lower than that of *minifundistas.*[72] A similar conclusion was reached in the Carirí where sharecroppers are better paid than wage-labourers but are worse off than tenant farmers or resident cowhands (*vaqueiros*).[73]

As a general estimate it seems that in the interior of North-East Brazil approximately one-third to one-half of sharecroppers' produce is consumed by the labourer and his family ('subsistence' production or 'autoconsumption').[74] This may vary according to individual contracts, to the size of establishment or even the general area. Some studies indicate a higher level of autoconsumption for particular regions: examples include the valley of the Banabuiú, Ceará (45%)[75] and the valley of the Acaraú (50%).[76] With certain staple food crops the level of subsistence consumption tends to be especially high, 80% of beans, manioc and other vegetables in the Jaguaribe valley, for example.[77] The same survey informs us that cotton (75%), caju (98%) and carnauba (90%) are mainly commercialized.

Although there is a high level of subsistence consumption, there exists a sufficient amount of direct market contact to allow us to place sharecropping in the category of 'semi-subsistence' agriculture, that is, farming which is devoted primarily towards autoconsumption and paying the landlord in kind, but a proportion of whose surplus also provides some sort of cash income for the *parceiro*. There is a conspicuous lack of reliable evidence on how much of his income the sharecropper receives in his hand free of any further obligations towards the landlord. The size of such an income, as well as any cash reserve, would clearly be important factors in determining how vulnerable the *parceiro* is during a time of need such as a drought when he could not necessarily count on the landowner's help to tide him over.

A sharecropper is fairly heavily dependent on his stock of subsistence crops to see him through to the next harvest, but it is unlikely that a family will be able to go through a whole year without needing at least some cash for other purchases or emergencies. Wage-labour may provide a useful supplement to a *parceiro*'s income. Unless there is a special demand for such labour which keeps rates up, as in the sugar *engenhos* of the Carirí,[78] it is normally only used as a last resort by sharecroppers. Wage working in Ceará was calculated as representing less than half the value of a sharecropper's labour on his own land, and sharecroppers avoided it if they had the chance to till their own plots instead.[79]

Where subsistence agriculture has been replaced by wage-labour in the *sertão* the effects on the rural population do not appear to have been beneficial. In Bahia, for example, sisal was introduced as an export crop and regarded favourably at first because it freed the population from their dependence on traditional agriculture. However, the substitution of wage-payments for subsistence farming caused a severe drop in the population's standard of living, including a marked increase in the incidence of malnutrition.[80]

The *parceiro* may also earn money by selling his share of the commercial crops, usually cotton and perhaps a few staple foods. Such benefits are substantially reduced in effect, however, for the sharecropper faces monopolies at every turn which reduce his bargaining power. When it comes to selling, the sharecropper has three alternatives: he may deal with his landlord, he could sell to a local *bodegueiro* or a travelling *comerciante* may take his produce away by lorry. Quite often these categories overlap, reducing still further the farmer's choice of market. The village shopkeeper, for example, may also be a landowner or a lorry-owning middleman collecting from many small producers. These intermediaries then sell to wholesalers (*atacadistas*) or even directly to the cotton gin. Commercialization of agricultural produce in the North-East is characterized by a long chain of intermediaries between the farmer and the purchasing public, each subtracting his profit margin. This helps to drive up prices for the end product and hold them down for the producers of the raw materials.[81] These basic features apply to agricultural marketing in the *sertão* whether cotton, fruits or vegetables are involved.

The dependence of the small-scale producer is typified as much by the low price he receives for his crop as by the high price he is obliged to pay for his inputs. This arises from the fact that because he has few assets and storage facilities, the poorer farmer is obliged to sell his crops when prices are lowest. To repeat, *venda na folha* is common still and may involve prices well below the subsequent market value.[82] This is typical not just of sales to the landlord but of dealings with other middlemen or *bodegueiros*, who advance supplies from their own stores on the same basis before the harvest.[83] Payment in cash is frequently replaced by payment in merchandise at prices above those which prevail locally (*escambo*).[84] Although this situation has improved with the extension of agricultural credit, *venda na folha* and other forms of price-cutting which prejudice the small farmer are still widely used, especially whenever there is a reduction in demand by the gins.[85]

Clearly, then, these restrictions on the bargaining power of *parceiros*, as well as on that of other small farmers, are bound to affect their susceptibility to the drought. This is particularly significant for the sharecropper because, by the very nature of his tenure arrangements, his contact with the market is already minimized in comparison with other occupational groups. These additional pressures limit the cash reserves which he can build up and, as a consequence, restrict his ability to diversify his assets as protection against the drought.

Discussion so far has been focussed principally on how the sharecropper is exposed to the effects of the drought by the landownership structure, the system of tenure arrangements and the process of commercialization. Much of what has been said applies equally to the smallholder or *minifundista* who also tends to be badly affected by the *sêca*. Property owners made up 19.3% of the rural labour force but constituted 31% of work front members in 1970 (see table 3). 85% of work front recruits come from properties of under 10 hectares, a fact which reflects the considerable pressures that small establishments are put under. *Minifúndios* are invariably more densely populated than larger holdings[86] and their resources have to be used more intensively in order to support dependents. A recent survey found that the product per hectare in the *sertão* on properties of less than 10 hectares is five times that on holdings of over 500 hectares. Conversely, the product per head on *latifúndios* is five times greater than that on the smallest establishments.[87] Nominally free, fragmented *minifúndios* are often dependent economically and politically upon large landowners and merchants. Like *parceiros* their relative lack of assets obliges them to commercialize their produce with larger landowners and middlemen who command a superior bargaining position.

During a crisis, however, *minifundistas* are generally in a better position to resist the drought than *parceiros*. They are not liable to be expelled by a landlord, nor do they suffer their land being invaded involuntarily by foraging cattle which devour the cotton and vegetable crop. They are more likely to have their own few head of cattle which can be sold in the event of a *sêca*; the land itself is another asset which may be sold as a last resort. Evidence of land sales during the 1970 drought suggests that *minifundistas* sold property and other capital goods disproportionately more than large landowners.[88] Most small properties, however, do not have the facility to decapitalize in the face of an emergency, either by selling cattle or land, and have to concentrate on maximizing their returns on crop cultivation. Using his family's labour and putting all his efforts into planting and replanting cotton as well as staple foods, the smallholder is in a somewhat better position to resist the drought than the *parceiro*. As we have seen, his real income is higher, his position more secure and less subject to the arbitrary decisions of landlords.

Institutionalized credit is one means by which the vicious circle of debt dependence may be broken. Unfortunately, those most vulnerable to the drought, because of their economically weak position, do not receive credit assistance in proportion to their needs. It has already been shown that larger holdings oriented towards extensive cattle-rearing are relatively cushioned from the drought for several reasons. Their position is further strengthened by both emergency credits and by normal agricultural loans, which have tended to ignore the urgent needs of the cotton and staple food-crop sectors, precisely those from which most *flagelados* originate.

Agricultural credit in the North-East has always been monopolized by larger landowners by virtue of the collateral they are able to offer banks. The Bank of Brazil (BB), for example, has always favoured lending to cattle enterprises rather than for the production of food crops since the collection and sale of animals in the event of default is much easier than if land as a guarantee has to be sold. The market for cattle is considered more secure and less liable to fluctuations than food crops. In normal years, therefore, properties of over 100 hectares tend to receive far larger bank loans than smaller properties.[89]

The bias towards cattle rather than perennial or food crops was reflected in the distribution of emergency credit during the 1970 drought. The CRAN programme (Special Credit for the Recuperation of North-East Agriculture) was authorized by the Central Bank in March 1971 and replaced the CEN.[90] CRAN was a one-year scheme which was intended specifically to finance, through the BB, BNB and BNCC (*Banco Nacional de Crédito Cooperativista*), special loans to medium and small producers with an annual production of up to 100 times the minimum wage in the case of the former and 1,000 times in the case of the latter. These limits could be exceeded in special cases where perennial crops (mainly cotton) and cattle (chiefly pasture-formation) were concerned. These exceptions, however, subsequently came to make the rule and dominated lending activities (see table 12). CRAN was replaced and absorbed by PECRO (Special Programme of Guided Rural Credit) in June 1971. The emphasis given to PECRO was more long term (1971–4) and on overall agricultural development. PECRO- and CRAN-subsidized credit together accounted for 37% of rural lending in the North-East by the BB in 1971. Credit was concentrated in improvements (48%), including pasture formation (25%), reservoirs (12%) and land clearance (7%) (see table 12).

The respective shares of cattle and cotton may have been in proportion to their contributions to regional agricultural production approximately

TABLE 12. *PECRO and CRAN special credit for North-Eastern agriculture, 1971 (in Cr$ millions)**

	Value		*% of total*	
Formation of perennial cotton	33.7		10.4	
Other perennial crop formation	23.3		7.2	
Acquisition of livestock	2.5		0.8	
Improvements	155.8		48.0	
Storage structures		3.5		1.1
Land clearance		22.5		6.9
Processing install.		3.6		1.1
Reservoirs, irrig.		38.3		11.8
Pasture formation		82.8		25.5
Rural residences		5.1		1.6
Machinery	7.3		2.3	
Agricultural machinery		3.5		1.1
Traction animals		3.8		1.2
Other improvements and machinery	100.1		30.8	
Other†	2.6		0.8	
TOTAL	325.3		100.0	

Source: Based on data from Bank of Brazil, 2nd Region.
*Both programmes were created in March 1971 by Central Bank Resolutions 175 and 181 respectively. PECRO covered the North as well as North-East but the latter accounted for 80% of loans. Land purchases under PECRO are not included in the data.
†Rural electrification, re-forestation, hatcheries, protection and restoration of soils, vehicles.

(25% and 9%), but livestock's share of emergency credits seems to have been disproportionately large in relation to losses incurred during the *sêca*. Cattle production actually increased during the drought year, as we have seen (in table 7), while cotton production fell by as much as 87% in drought-affected areas, although the overall decrease in cotton production may have been nearer 40%.[91] Staple food crops fare particularly badly during a drought. At the first sign of a *sêca* or abnormally dry period banks withdraw their loans for fear of losing their money and renew them only when the first rains arrive, which may be several weeks after planting starts. This is true of both the BB and BNB which together account for virtually all institutionalized lending to agriculture in the *sertão*.[92] Subsistence crop production is concentrated on smaller properties, precisely those which stand the smallest chance of maintaining the pre-drought level of credit.[93] In 1972 basic staple foods (beans, manioc and corn) accounted for 19% of the total value of crop production in the North-East and 42% of the total cultivated area, but received only 19% of agricultural credit in that region.[94]

The PROTERRA credit programme,[95] which took over from PECRO and CRAN in 1972, continued this bias. The maintenance and acquisition of livestock was given preference, with PROTERRA cattle loans as a proportion of the total PROTERRA loans increasing from 45.7% in 1972 to 51.6% in 1973.[96] This is also reflected in the large size of cattle loans compared with those for crop production. Livestock loans went up by 27% over the period 1970–3 while loans for crops remained stable (see table 13). The share of livestock credit in total BB lending in the North-East rose from 25% to 33% from 1970 to 1975, although the figure for Brazil was a constant 21% over this period.[97]

Although PROTERRA was intended to provide subsidized credit (actually at negative interest rates) for investment, modern inputs and land purchases in North-Eastern agriculture, there are indications that it may not be serving this purpose. The short-term nature of loans under the plan conflicts directly with its manifest goals. Credit is given for a five-year period under present arrangements, which has the effect of favouring more traditional, extensive methods of cattle rearing. Short-term, speculative investment may be the result of this policy rather than the longer-term commitment towards modernization implied by PROTERRA. It is conceivable that existing pastures could be depleted or even cotton- and food-producing areas converted to cattle grazing in order to obtain quick returns. If this were to happen on a large scale, the displacement of sharecroppers and their families already typical of drought-hit regions may become far more widespread in the *sertão*.

To conclude the discussion on how the rural structure of the North-East affects the vulnerability to the drought of certain groups more than others, a word on current trends is in order. The changing pattern of landownership will be looked at, along with modifications in the occupational structure of rural areas in the *sertão*. These factors are likely to have an important bearing on the size of groups at risk to the *sêca* as well as their degree of vulnerability.

Over the past thirty years in North-East Brazil food production has expanded in step with increased urban demands, chiefly by means of a proportional increase in the area under cultivation rather than through higher productivity.[98] In fact, production of food crops has increased twice as fast as industrial crops.[99] Two seemingly opposed schools of thought exist which try to explain how the rural population was mobilized

TABLE 13. *Bank of Brazil agricultural loans, Brazil and North-East, 1968–73**

	Crops				*Livestock*			
	Brazil		N.E.		Brazil		N.E.	
Years	No. (000s)	Average value	No. (000s)	Average value	No. (000s)	Average value	No. (000s)	Average value
1968	473.3	10.7	65.8	9.3	102.9	11.5	15.2	12.9
1969	448.5	10.9	59.0	9.9	118.1	10.9	18.1	12.3
1970	491.1	11.2	63.0	8.6	121.8	10.0	15.1	11.2
1971	560.4	11.2	111.6	6.9	150.4	11.7	28.4	13.4
1972	599.6	13.7	105.3	8.7	165.0	12.4	31.0	13.2
1973	643.3	13.6	104.3	8.6	168.9	14.8	32.5	14.2

Source: Based on data from Bank of Brazil, 2nd Region.
*1972 Cr$ thousands.

to supply this increased demand. The 'pre-capitalist' model views the maintenance of non-capitalist relations of production as a prerequisite to capitalist expansion in the North-East which releases an agricultural surplus for the urban sector. This analysis sees agricultural production of foodstuffs as an almost completely subsistence activity which expands in relation to the needs of its members rather than in response to market pressures. Any surplus provides the urban food supply, and the fact that they coincide is fortuitous.[100]

An alternative view, the 'capitalist' model, holds that this post-war increase in agricultural output is a direct response of producers in the countryside to income opportunities generated by increased urban demand, facilitated by better communications and the penetration of an increasingly rationalized peasant marketing system. Farmers and merchants alike act in accordance with market incentives.[101] Another argument, compatible to some extent with both schools of thought, attributes the higher output to reduced underemployment of labour and higher productivity per capita resulting both from internal migration and net loss of labour from the North-East as a whole.[102] This analysis is more similar to the 'capitalist' model and differs fundamentally from the 'pre-capitalist' in assuming a much greater degree of commercialization and price motivation on the part of small producers.[103]

While these analyses differ on the degree of commercialization of agriculture that exists in the countryside and on the nature of small producer motivation, all agree that some contact with the market, whether directly or indirectly, is involved. There is, in other words, more overlap between these apparently contrasting viewpoints than some critics are prepared to admit. The above discussion of sharecroppers and smallholders in relation to the drought labelled this sector 'semi-subsistence' in recognition of this fact. That is, both *parceiros* and *minifundistas* devote themselves principally to satisfying subsistence needs but are linked to the market in varying degrees by dealings with landowners and middlemen. The more direct contact of large landowners with the market economy, mainly through the commercialization of cattle and cotton, was found to be a major factor shielding them from the harmful effects of the drought. The relative isolation of sharecroppers and, to a lesser extent, smallholders, from the market means that these groups suffer disproportionately by having insufficient resources to fall back on in terms of cash and other liquid assets. Any expansion or contraction of this 'semi-subsistence' sector will, therefore, have a fundamental bearing on the human impact of the drought.

One clear trend which has emerged from an analysis of Agricultural Census data is the polarization in the distribution of landholdings. As table

14 shows, over the period 1940–70 in the *sertão* properties below 10 hectares accounted for 7.55% of the increase in cultivated area, while establishments of over 100 hectares were responsible for 60% of this increase.[104] There has been a large increase in the number of smaller properties, those of under 10 hectares rising from 36% of the total in 1940 to 58% in 1970. Over the same time average property size in this category has fallen from 5.41 hectares to 3.24 hectares. At the other end of the scale the proportion of establishments over 100 hectares has dropped from 22% to 9%, while their share of the total cultivated area has remained constant at around 75% (see table 9). Such evidence suggests that there is taking place a fragmentation of *minifúndios* in the *sertão* through inheritance and land sales which is comparable with trends for the North-East as a whole.[105] The statistics also indicate that, as far as larger establishments are concerned, increased concentration of *latifúndios* into fewer hands and, more significantly perhaps, into fewer economic units, is becoming increasingly evident.

Such trends can only be encouraged by current government policy over agricultural credit. In their desire to encourage the spread of 'agro-industry' Brazilian planners have made available large sums of subsidized credit through the banks to farmers wishing to expand extensive cattle-raising. As we have seen, this has happened under PROTERRA, but it is also taking place under SUDENE's 34/18 fiscal incentive scheme in which over 90% of

TABLE 14. *% increases in land areas accounted for by different property sizes, sertão, 1940–70*

	Size of landholdings (ha)					
States	0–1.9	2–9.9	10–99.9	100–999.9	1,000+	Total
Pernambuco	0.82	2.80	30.29	45.21	20.88	100.0
Paraíba	1.24	6.44	32.10	23.42	36.80	100.0
Rio Grande do Norte	1.16	10.88	28.08	18.29	41.59	100.0
Ceará	0.48	7.98	37.00	30.57	23.97	100.0
Piauí	1.48	2.95	21.69	40.70	33.18	100.0
Bahia	1.17	8.67	42.62	40.27	7.27	100.0
Sergipe	0.44	2.73	34.50	44.17	18.16	100.0
Alagoas	2.52	7.80	34.90	42.11	12.67	100.0
TOTAL	1.10	6.45	32.98	38.00	21.47	100.0

Source: Calculated from IBGE, *Agricultural Census*, 1940 and 1970.

loans are going into livestock enterprises rather than other forms of rural activity.[106] Although increases in livestock production and exports produce impressive economic statistics, such productivity could be prejudicial to the labouring population affected by this change in land-use. Cattle-rearing uses relatively little labour, one of the main reasons for the limited human impact of the drought until the nineteenth century, when staple foods and cotton became important crops in the *sertão*. The spread of livestock enterprises is, therefore, likely to displace labour, especially if lands currently used for subsistence food and cotton are taken over for pasture. This has happened in marginal cocoa-producing areas of Bahia, in the *agreste* of Pernambuco and in the upland *serras* of Ceará.[107]

It therefore seems probable that a similar process is liable to take place in the *sertão* proper. Unless such displaced labourers migrate from the area in search of permanent rural or urban employment, they would be forced to join the ranks of temporary wage-labourers in rural areas which, according to past experience, tends to lower their real standard of living. In theory this might at least perpetuate if not increase the harmful effects of the drought upon this section of the population. If the effect of credits to livestock is as described, they will only be detectable in the statistics in the long run, but their impact will be significant.

Another discernible trend is the growth of that sector of the rural population classed as the head of the household together with his unpaid family labour. Some observers have taken this development as the consequence of the inability of the urban economy to absorb excess rural population.[108] In the *sertão*, the unremunerated family labour sector has expanded from constituting 62% of the rural labour force in 1960 to no less than 83% in 1970 (see table 11).[109] This increase has been at the expense of other occupational groups such as hired labour and sharecroppers. The definition of sharecropper adopted by the IBGE in table 11 is useful for present purposes since it includes only those *parceiros* not 'responsible' for administering the land. Sharecroppers with such responsibility are included in the family labour group. The former type of sharecropper (*o não responsavel*) was discovered by the 1970 drought survey to have the least security of tenure and to be the most liable to be affected. During the decade 1960–70 unpaid family labour almost doubled from 1,200,000 to 2,200,000 in the *sertão*.[110] Subordinate sharecroppers experienced a small decrease in numbers from 127,000 to 119,000, although as a proportion of the total this group fell from 6.24% to 4.58% over the same period (see table 11). The decrease in the number of subordinate *parceiros* could be indicative of a smaller population at risk to the drought. On the other hand, the growth of the family labour sector on *minifúndios* which have gradually become more fragmented over the years could increase further still the pressures on their limited resources. Furthermore if, as some have suggested, the excess of rural population left after emigration tends to become concentrated in subsistence farming, their vulnerability during a drought will undoubtedly increase because contact with the market seems to act as a cushion against the drought and vice versa.

This emphasis on the social and economic roots of the so-called 'drought problem' stands in direct contrast to the official preoccupation of anti-drought agencies such as DNOCS with poor rainfall as the direct cause of human suffering during extended dry periods. It became clear during the first half of this century that DNOCS's concern with reservoir-building in response to the *sêca* did little to diminish the susceptibility of the mass of poorer *sertanejos* to the drought. This was due in part to the fact that reservoir-building was not accompanied by the construction of irrigation canals that might have enabled a larger proportion of the rural population to benefit from the enormous volume of water that was stored. As it was, only nearby landowners who grazed their cattle on the water's edge and a handful of small farmers who practised *vazante* agriculture were protected from the drought by the reservoirs.

Irrigation is the latest anti-drought strategy and is designed to correct this imbalance by helping to alleviate problems of rural unemployment and migration associated with the *sêca*. The following chapters will examine in detail, using three case studies of irrigation projects, whether this goal is being met.

3. The origins and state of irrigation in North-East Brazil

For over a century the drought in North-East Brazil has been regarded as one of the central causes of the region's socio-economic backwardness. Yet it was not until the late 1960s that irrigation as an anti-drought strategy was seriously considered by the government, despite the fact that it had been used for thousands of years in many semi-arid parts of the world. Indeed, small private irrigation systems have long been a feature of the Brazilian *sertão*.[1] The *idea* of using irrigation to fight the drought had been mooted as early as the 1890s when the first large-scale reservoir, the enormous Açude Quixadá in Ceará, was constructed. With a 50-kilometre network of canals, it had the capacity to irrigate several thousand hectares. However, no plans were made to provide facilities for crop production and nearby landowners used the reservoir for watering their cattle pasture.

The serious drought of 1930–2 finally provided the impetus needed to get seven small irrigation projects started, but they had no significant impact against the drought. In 1932 Getulio Vargas, aided by a north-easterner, José Américo de Almeida, as his Minister of Public Works, increased federal spending on the region to 10% of total government revenue. In the same year work was started on the Lima Campos project (Ceará) followed by São Gonçalo (Paraíba) and Joaquim Távora (Ceará) in 1933. This first generation of irrigation schemes came to include Condado (Paraíba), São Francisco (Pernambuco), Itans (Rio Grande do Norte) and Santo Antônio de Russas (Ceará). Progress was exceedingly slow, however, and by 1941 only 500 hectares of land had been irrigated.[2]

São Gonçalo seems to have been the most successful of these early projects and illustrates very well the limitations of the old system. An agricultural station (*posto agrícola*) was established here by IFOCS in 1941, at which time only 118 hectares had been irrigated. The 1942 drought sent hundreds of families from surrounding areas to São Gonçalo seeking shelter. Private landowners, who possessed most of the land and bought water from IFOCS very cheaply, took advantage of the influx to expand their labour forces and the area under cultivation to over 1,000 hectares by the end of 1942. IFOCS, which also owned and rented out an area of farmland, placed a further 400 hectares of land under irrigated production.[3]

The benefits of this type of irrigation scheme were spread very unevenly, however. Only a small proportion of the land was actually owned by

IFOCS and rented out to smaller cultivators (16%). Most of the project area (84%) belonged to private landowners. Over 60% of the land was concentrated in properties of over 40 hectares, which is considered to be large for a relatively fertile, densely populated valley.[4] A high proportion of the larger landowners were totally absentee, leaving their properties in the care of administrators and dividing them into small plots for tenant farmers and sharecroppers. The latter two groups were found to have a real income equivalent to just over half the official minimum salary in 1968. On average, large properties (50 hectares or more) enjoyed an income fourteen times that of medium-sized holdings (10–50 hectares) and twenty-five times that of small properties (under 10 hectares).[5]

A comprehensive survey of the area concluded that only the largest landlords made a 'reasonable' living from the land, although other sources of income are not mentioned. The 0.1–5.0 hectare category accounted for 32% of properties but only 3% of the area. As a consequence of this intense fragmentation, three-quarters of this group of *minifundistas* earned less than the minimum salary. The proportion of sharecroppers earning less than the official minimum was 63%, and 93% for *moradores.*[6] Family incomes were kept low, the study concluded, by the high proportion of dependents and lack of employment opportunities.

For almost thirty years these few hundred hectares of largely private land served by DNOCS-built canals constituted the entire public effort towards irrigated agriculture in the *sertão*. Only in the aftermath of the disastrous 1958 drought was there a change of strategy away from reservoir-building. SUDENE subsequently started work on two projects in the São Francisco and Jaguaribe valleys in the 1960s but it was not until the end of the decade that irrigation's place in regional development plans became assured.[7] GEIDA, formed in 1968, was charged with undertaking a comprehensive study of irrigation possibilities for the whole country. In this year the Minister of the Interior, General Albuquerque Lima, tried to promote more redistribution and placed greater emphasis on agriculture and irrigation. The various North-Eastern agencies were gradually centralized under MINTER, and DNOCS lost its autonomy in 1968. During the period 1968–70 SUDENE and SUVALE received a more or less constant level of funding from the federal government while DNOCS's support dropped by almost 50% (see table 15). This was, no doubt, due at least in part to the better record of the former two agencies in contrast to DNOCS's reputation for misdirecting funds, acquired during and after the 1958 drought. However, under the GEIDA proposals, investment into irrigation schemes in the North-East (Region A), over which DNOCS has almost exclusive control, will be around Cr$1,232 million. This compares with

TABLE 15. *Federal funding for irrigation to North-Eastern agencies, 1966–70 (in Cr$ millions)*

	MINTER			SUDENE		SUVALE		DNOCS	
Years	Total	Ag.	Irr.	Ag.	Irr.	Ag.	Irr.	Ag.	Irr.
1966	303.4	23.1	0.6	10.1	–	4.9	0.6	*	*
1967	379.9	36.6	8.3	22.4	–	11.9	8.3	*	*
1968	618.9	108.1	62.2	25.5	3.1	10.1	11.5	45.5	39.5
1969	701.9	113.9	75.5	30.4	11.3	16.9	13.5	41.6	38.5
1970	557.8	78.1	50.6	27.6	11.1	12.5	11.1	26.7	23.8

Source: GEIDA, *Plano Nacional de Irrigação, Diagnóstico Preliminar*, Brasilia, 1970, table B5.1, p. B5.11.
Notes: *DNOCS was not in MINTER.
Ag. Agriculture, including livestock.
Irr. Irrigation

investment into the São Francisco valley (Region B) under the direction of CODEVASF which amounts to Cr$2,005 million until 1980. These sums, at 1971 values (when US$1 = Cr$4.20), are in proportion to the areas of land to be irrigated.[8] Irrigation seems assured of a steady source of financing at least until the end of the 1970s, which is the period covered by the initial *PPI* proposals.

The objectives of irrigation strategy regarding its contribution to Brazilian economic development are quite specific and may be dealt with under the headings of employment, income and agricultural production. GEIDA hopes to create no less than 115,000 jobs directly on projects in the North-East and São Francisco valley by 1980, on the assumption that schemes will be fully operational. These jobs would benefit a total of 350,000 people including dependents. In addition, 230,000 jobs would be created indirectly, benefitting another 700,000.[9] The total number of people affected is therefore estimated at over one million by 1980, out of a total rural population in the North-East estimated to be nineteen million by then. Labour absorption and the stabilization of employment is perhaps the main goal of irrigation. The heavy incidence of unemployment and underemployment during normal times, exacerbated by the drought which throws thousands of small producers off the land, makes the creation of permanent jobs especially relevant in the *sertão*. Irrigation projects in the Centre-South of Brazil (Regions C and D) will be more important as a source of supply for vegetables and fruit in order to eliminate seasonal price variations.[10]

GEIDA also predicts a large increase in the incomes of rural population which benefits directly from irrigation projects, Cr$2,100 (US$360) per annum per capita, as opposed to the estimated average of Cr$583 (US$100) in 1980. The total annual income generated per person directly employed is put at Cr$6,640 (US$1,581) to Cr$11,200 (US$2,666) in contrast to Cr$2,940 (US$700) outside of irrigated agriculture.[11] This is set against investment costs estimated at US$2,000–3,500 per hectare, or US$4,500–7,000 per worker, on the assumption that 1.8 to 2.5 hectares of irrigated land are necessary to create one permanent job.[12]

The *PPI* plans suggest that one of the results of this employment and income-generating activity will be to help slow down the rural–urban exodus in the North-East and thus diminish the pressures on a heavily inflated tertiary sector in the cities.[13] Another will be to increase agricultural production both for the domestic market and for export overseas. A substantial income will be needed, GEIDA maintains, in order to justify the heavy infrastructural and other investment in irrigation.

GEIDA's crop production plans are based on calculations of probable demand potential for a number of industrial and food crops including cotton, citrus fruits and vegetables. The profitability of irrigation schemes depends to a large extent, GEIDA maintains, on sales of high-value crops. Projecting ahead to 1980, when projects would be operating at a minimum 10% rate of return, output will be very diversified. In the North-East citrus fruits head the production list (630,000 tons per annum), followed by assorted vegetables (383,000), tomatoes (302,000), cotton (254,000), potatoes (225,000) and other fruits (102,000). The rest, in order of importance, include melon, fodder crops, grapes, bananas, sunflowers, onions, peanuts, beans, wheat, pineapple, garlic, rice, corn and sorghum.[14] These optimistic forecasts extend to predicting exports of several hundred thousand tons of fruit to Europe and the USA by 1976.[15] The total value of exports from PPI projects is expected to reach US$12.6 million for high-value fruit and vegetables, and US$34.5 million for traditional crops.[16] Such an increase in agricultural exports is seen as an important contribution towards the balance of payments.[17]

In achieving these three objectives, those of increasing agricultural employment, incomes and production, GEIDA forecasts that valuable side-effects will follow. Firstly, irrigation farmers will acquire technical skills, organizational capacity and the ability to 'cooperate positively' with their fellow producers. In addition, it is hoped that they will develop 'more rational behaviour in the sense of maximizing their income'.[18] Presumably this is based on the assumption that rural workers are at present non-rational beings, content to earn less-than-maximum incomes. A second side-effect will be for technological diffusion to take place from projects to non-irrigated agriculture. In this way, it is suggested, the transfer of ideas and production techniques will boost incomes in surrounding areas as well.[19]

When the GEIDA report was published in 1971 Brazil had 520,000 hectares of irrigated land, 340,000 of which consisted of rice-fields in Rio Grande do Sul. Federal projects in the 'drought polygon' accounted for under 10,000 hectares and many schemes were in a precarious state. During 1965–70 DNOCS had officially started work on a handful of irrigation projects but only a few dozen families were recruited to cultivate less than 1,000 hectares of irrigated land. The GEIDA plan incorporated these projects into the regional plan and tried to systematize their execution. GEIDA envisages irrigating 368,000 hectares in Brazil by 1980 in sixty-four 'economically viable' projects (defined as yielding a minimum rate of return of 10% on capital investment). The North-East and São Francisco valley (Regions A and B) will account for 195,000 hectares. DNOCS has jurisdiction over thirty-six projects, thirty-one in Region A and five in Region B, covering 78,000 hectares. CODEVASF will operate seventeen projects in the São Francisco valley which together account for 112,000 hectares. Finally, DNOS (*Departamento Nacional de Obras de Saneamento*) has charge of three schemes and some 5,000 hectares, which fall in both regions.[20] In addition, DNOS is completely responsible for irrigation projects in the Centre and South (Regions C and D), another 172,000 hectares organized into just eight schemes. Overall then, of the three agencies involved, DNOCS has the largest number of projects (thirty-six out of sixty-four) but the smallest area of irrigated land (73,000 hectares of 368,000).

The situation over goals is a little confusing, however, since DNOCS's own published figures differ from those in the GEIDA study. The North-Eastern agency has a rather more ambitious target of 100,700 irrigated hectares by 1980, as well as a further 274,000 hectares of *caatinga* incorporated to combine with the agricultural calendar of irrigated areas, providing pastures for grazing cattle in winter months principally.[21] DNOCS also hopes to have absorbed 21,300 families onto irrigated plots by the end of the decade.[22] Perhaps these differences between DNOCS and GEIDA goals represent additional targets that the former agency has set itself, over and above *PPI* proposals.

An additional facet of irrigation plans is the concept of 'priority areas'. This idea was originally developed in Brazil by President Costa e Silva in 1967, the same year in which the various North-Eastern agencies became subordinated administratively to the Ministry of the Interior. During one particular week, when the President moved the site of the federal capital to Recife in order to focus attention on the region's problems, a new plan was formulated. The 'Co-ordinated Action of the Federal Government in the North-East' was a series of 'principal measures and priority projects' of the administration. The Ministries of Planning, Housing, Agriculture and the Interior established a set of *áreas prioritárias* upon which efforts would be concentrated. These included the humid valleys and uplands of the interior, the coastal table-lands, the *cerrados* of Minas Gerais and the wetter lands of Maranhão.[23] This has been interpreted as a means by which the government could demonstrate its interest in the region without being required to release funds for specific projects.[24] However, the systematization of plans for the region, of which the 'priority area' programme was a part, may have been motivated at least partly by the fact that the World Bank, the probable main external source of finance for agricultural projects in the North-East, wished to have evidence in front of it that planning progress was being made. Ultimately, loans would only be given after specific projects had been vetted by the IBRD.[25]

These priorities were formalized into the POLONORDESTE strategy in 1974 under the Second National Development Plan. Similarly, DNOCS itself has established 'priority projects' concentrated into five areas: the valleys of the rivers Parnaíba (Piauí), Jaguaribe (Ceará), Piranhas–Açú (Paraíba), Ceará Mirim (Rio Grande do Norte) and Moxotó–Gorotuba (Pernambuco), which encompass eleven 'priority projects'.[26] Two of these schemes, São Gonçalo in the Piranhas valley and Morada Nova in the Jaguaribe valley, are included in my three case studies.

By late 1975 only a small fraction of DNOCS's objectives had been realized. Its 1974 goal of 36,000 hectares in operation had long been forgotten.[27] Of its 100,000 hectare target for 1980 only 9,000 had been irrigated. Neither was there any indication that the pace was suddenly going to be stepped up to 20,000–25,000 hectares per annum in operation, as GEIDA envisaged. Of 21,000 families DNOCS hopes to absorb by the end of the decade only 900 have so far been accommodated. CODEVASF's record is worse still with only two of its planned seventeen projects functioning and 150 families installed.[28]

The administration of DNOCS's activities in the field of irrigation is divided into three levels: central, regional and individual projects. Although the agency's representation is therefore decentralized, quite the opposite seems to be true of decision-making. All major policies are formulated and decisions made by the Directive Council (*Conselho Diretor*) which is made up of a Director General and six members appointed directly by the Minister of the Interior. Subordinate to the Directive Council and responsible for coordinating the various divisions is the General Directorate (*Diretoria Geral*). The General Directorate embraces four divisions (planning, operations, administration and personnel) as well as the four regional directorates. The *diretorias regionais* each have a central office, situated in Fortaleza, Recife, Teresina and Salvador, while the central administration occupies a large, modern building in Fortaleza. The regional offices are each subdivided into several sections, responsible for project study, engineering, social work, administration, personnel and pisciculture. Individual irrigation project managers are directly responsible to their respective regional director.

Generally speaking, regional offices seem to have little autonomy to make decisions themselves without prior consultation with and authorization from the central administration in Fortaleza. Similarly, project managements have a limited degree of freedom to act without first obtaining permission from regional head-office which may, in turn, have to receive the go-ahead from central administration. On the one hand this system has the apparent advantage of controlling the application of resources in the light of DNOCS's bad reputation derived from the *industrias da sêca*. Although this particular phase during the late 1950s was the only stain on an otherwise clean record, DNOCS personnel are still very anxious to avoid even the suspicion that there could be a recurrence of corrupt practices within the organization. On the other hand, the numerous stages involved in the decision-making process often lead to unnecessary delays and conspicuous over-bureaucratization.

In common with the overall administrative structure of DNOCS, the internal organization of individual projects is very hierarchical. There are three types of project administration, depending on the size of project, but all seem to share one common, fundamental characteristic. On all three case-study projects there exists a visible and profound split between the management staff (*gerência*) and the irrigation farmers (*colonos* or *irrigantes*). There is little exchange of ideas and no collective decision-making involving *gerência* and *irrigantes* working together. Technical and administrative functions are housed in the main office or *gerência*, as it is known, and decisions regarding these matters are imposed exclusively from the top downwards. The opinions of the farmers themselves regarding agricultural production, system of credit, payment of incomes and so on, aspects which directly concerned them, were, it seemed to me, rarely invited by the *técnicos*. The organizational structure simply does not cater for this. Complaints by farmers regarding their lack of say in the way the projects were run were often matched by management's dissatisfaction with arbitrary decision-making on the part of regional and central offices unfamiliar with local conditions.

In theory the interests of *colonos* are represented by the project cooperative. All farmers are automatically members. The cooperative is run by a 'deliberative council', which consists of members from the *gerência*, the technical staff in charge of cooperative administration and the farmers. Only one elected representative from the *colonos* sits on the council, whatever the size of the irrigation scheme. In practice the cooperative concerns itself exclusively with distributing credit, selling the farmers' produce and keeping charge of the accounts. The farmers have little or no effective say in how the project operates. Group meetings of the *colonos* with the *gerência* are rare and, when they do take place, criticism of existing procedures is certainly not encouraged. Most farmers are unwilling to speak up in meetings. The tendency is, rather, for the acceptance of decisions handed down from top level.

As might be expected, the distance between project administration and farmers is reinforced by a total contrast in life-styles. The fact that higher level *técnicos* have completely different backgrounds, live in modern, large houses well away from the farmers, send their children to different schools and do not, as a consequence, mix socially at all, might seem an unnecessarily obvious point to make. It is, however, an important facet of day-to-day life which matches and strengthens the divisions that already exist in working relationships.

The task of planning how an irrigation project will be executed is not officially in the hands of DNOCS itself. A number of private companies have been contracted to undertake feasibility studies of the various river basins considered by DNOCS to be suitable sites in which to locate projects. These consultants are both European and Brazilian: Tahal Consulting Ltd (Israel), SCET (France), SIRAC, Hidroservice and Tecnosolo (Brazil) are a few examples. Firms are contracted on the basis of tenders submitted by competing companies to the Ministry of the Interior. Contracts are much sought after and, by all accounts, very lucrative. It was not possible to obtain precise information on this but one *técnico* informed me, for example, that the viability study for the Vaza Barris project in Bahia cost the Brazilian government over one million dollars. Whatever the exact amounts, there exists the possibility that strong efforts will be made to 'prove' the economic viability of irrigation schemes on paper. It is in the interests of the company to do so because favourable cost-benefit ratios justify the high costs, both of the study and implementation of the project. This is especially so in the light of the GEIDA precondition that a project to be included in the *PPI* programme should have a rate of return of at least 10%. As far as DNOCS is concerned favourable studies serve to justify the long-term viability of irrigation as a development strategy and guarantee the flow of federal funds. The present research indicates that this practice has led to unjustifiably optimistic forecasts for irrigation which, as will be shown in later chapters, have not been matched by subsequent developments.

A viability study may be broadly divided into a pre-study phase and a planning stage. The first is taken up with assessing the general technical and socio-economic problems of the area, leading to a set of general recommendations. Such pre-studies invariably reaffirm the importance of irrigation to the valley in question if it is to 'progress' and if the living standards of the population are to be raised. No other strategies are ever suggested as serious alternatives. Once approved by the DNOCS central administration the feasibility study moves into the main planning phase, which consists of four stages. Firstly, the river basin's resources are surveyed, including climate, soils, hydraulics, engineering problems, agricultural potential and socio-economic features. The second and third stages pursue deeper studies of the same phenomena and broaden the scope to encompass market studies, labour absorption, cost-benefit analysis, the planning of engineering and colonization details. Finally, the executive project involves detailed mapping of the area, a comprehensive soil survey and plans for the successive stages of the irrigation project as it is evolved.[29]

With a comprehensive plan as its basis, work on the project itself is then started. The first task is expropriation of the privately owned land in the river valley. This process will be discussed in greater detail in the next chapter, but the point should be made here that expropriation is often hurried. A combination of political pressures from Brasilia and the desire of DNOCS to be seen to be making headway towards its official targets frequently speeds up the process, which exacerbates already existing hardships for the population forced to leave. DNOCS sometimes contracts special firms on its larger projects simply to make sure that families leave the project area by the deadline which has been set.

Irrigation schemes are located in valleys which already possess reservoirs built by DNOCS. The agency employs private engineering companies, contracted after competing tenders have been assessed, to clear and level the land as well as build the network of canals. Irrigation is principally by gravitational flow; farmers draw their water from tertiary canals, which serve each plot of land, using flexible plastic tubing (*sifão*). Construction companies are hired to build the project offices, staff houses, farmers' houses and garage and store facilities. On the larger projects, such as São Gonçalo and Morada Nova, *colonos*' houses are grouped together into several nuclei. The smaller schemes such as Sumé have a house adjacent to each plot. Sometimes existing houses, expropriated along with the land, are utilized in order to minimize costs.

In order to assess the performance of irrigation, three projects were chosen as case studies from the twelve which DNOCS had in operation during the time that field research was undertaken. Several factors influenced my choice of these specific irrigation schemes. Firstly, a rough cross-section of projects was required in terms of size (families absorbed and area irrigated). Secondly, it was felt that, for the purposes of evaluation, all case studies should have been in operation for several years so that significant achievements and trends could be identified. Finally, adequate documentation for all three projects was essential. This consisted primarily of information referring to the characteristics of the valleys before the projects were implemented, crucial to any comparative analysis. In one case this was provided by DNOCS (Sumé) but in the others (São Gonçalo and Morada Nova) the information was obtained from feasibility studies made by private consultants.

A potential methodological problem presented by any such selective sampling is, of course, the degree to which one can generalize from a few case studies. This problem, although uppermost in my mind throughout field research, is not thought to detract from the validity of general conclusions reached about irrigation projects in their present form. There are several reasons why this is so. Firstly, DNOCS adopts a similar strategy, with minor variations, for all its irrigation schemes, for example, in terms of expropriation and compensation, division of landholdings, commercialization, authority structure and method of recruitment. Secondly, the rural structures of the valleys upon which this system is imposed share common basic characteristics, such as the structure of property ownership, tenancy arrangements, system of crop commercialization and demographic pattern. As a result of these two facts, during field-work it was found that, again with certain exceptions, a pattern emerged on all three case studies with the same basic features. This was not fortuitous but the logical result of a similar DNOCS strategy for rural areas with basically similar socio-economic situations.

This is not to say, however, that significant methodological problems were not encountered. The prospect of trying to obtain information concerning the internal organization of irrigation projects was rather daunting at first. This was particularly true about delicate matters such as incomes and expropriation procedures. As it turned out, however, a combination of documentary evidence and information given by project personnel provided adequate data. At the level of face-to-face contact with DNOCS staff and irrigation farmers the degree of cooperation varied considerably. Generally speaking, the higher the official status of an interviewee the less likely was that person to commit himself or divulge information which could be said to do anything but bolster the image of the administration. Project directors would invariably produce optimistic estimates of the numbers of families absorbed, area irrigated and other achievements. Conversely, less flattering statistics on the number of people displaced from their homes by schemes would always be underestimated. Only one of the three directors openly offered constructive criticism of DNOCS tactics. At best, informants avoided giving embarrassing answers and simply withheld information. At worst, deliberate deception was attempted with incorrect answers on fundamental points which could not fail to have been known by those interviewed. However, because my own enquiry had official backing, the general response of higher office-holders was one of polite tolerance.

Consequently, of course, all information gathered from such sources had to be cross-checked. This was relatively easy to do using official documents, obtained after some persistence, as well as independent studies and interviews with lower-level employees. Gradually, all the facts fitted into a comprehensive picture. *Técnicos* lower down in the hierarchy such as social workers and agricultural technicians were particularly helpful in providing information. These tended to be people with more day-to-day contact with grass-roots problems of the project. Partly because of their greater experience and partly because they felt freer to divulge a whole series of facts which the higher administration was anxious to conceal, DNOCS staff at this level were extremely expansive.

The fact that two such contrasting reactions to my presence were to be found within the projects studied, especially in the two larger schemes, is itself significant. It is indicative of organizational structures in which there are considerable restrictions on the exchange of ideas and flow of information between the levels. As far as field-work in such a situation is concerned, it obviously had important repercussions. Apart from influencing the reactions of higher employees to my presence, it also conditioned the responses of the irrigation farmers. Being obviously an outsider, they usually thought at first that I was a DNOCS official on some supervisory mission. This made the *colonos* reluctant to communicate, especially if approached formally and/or accompanied by a project staff member (with the possible exception of certain social workers who did not have this effect on their reactions). Consequently, I attempted to clarify my independent position, keeping interviews as informal and as casual as possible. This tactic produced surprising results. Previously taciturn *colonos* voiced their opinions on the project, frequently with constructive suggestions on how agricultural production could be improved to produce what they felt would be better results. Much valuable information was obtained in this way concerning *colonos*' views on what was being achieved with irrigation. It provided a useful alternative picture to that portrayed by DNOCS.

Project directors never encouraged me to talk to the farmers directly. I got the strong impression that the exercise was not looked upon favourably by the *gerência*. When I was taken on conducted tours by higher officials it was always, as I subsequently learned, to the houses of better-off farmers and to the more attractive plots. Although this was perhaps understandable, it did not serve to present a balanced picture of what happens on an irrigation project. My own enquiries, assisted by lower-level staff, corrected this imbalance. To cite one example, on the Sumé project the director first took me to visit an irrigation farmer who had been particularly successful. His house boasted a TV set, refrigerator and electric sewing machine. The suggestion made to me that he was somehow typical did not stand up in the light of my own subsequent research, which showed that he was very much the exception to the rule.

Sumé is the smallest and longest established of the three projects studied. It covers 260 hectares of land in the Sucurú valley, Paraíba, about half of which are irrigated. It is fed by the Açude Sumé, constructed by DNOCS in the early 1960s, with a capacity of 37 million cubic metres.[30] The project is situated in the *município* of Sumé, adjacent to the town of the same name, which has a population of some 2,000. The general area, known as the Sertão dos Cariris Velhos after the indigenous tribe whose domain it once used to be, is one of the driest regions in the interior with a mean annual rainfall of 437 mm.[31] The surrounding countryside is typical *caatinga*: shrubs, trees and xerophytic vegetation used for open-range grazing of cattle. Typically, before the project was started, the valley provided a good deal of protection for the population despite the fact that the river was usually dry during the summer months. The land was cultivated by a fairly dense population of semi-subsistence farmers and larger landowners using traditional *sertão* techniques and producing the usual crops of aboreal cotton, beans and corn (*algodão*, *feijão*, *milho*).

The first families were installed in 1969 and successive intakes had brought the total number to twenty-six by late 1975. Sumé is considered to be an economically successful project which DNOCS takes pride in showing to visitors. It was because of this fact, rather than in spite of it, that Sumé was chosen as representing many of the features which the agency would like to see develop on its other projects and which it encourages. The official aim of DNOCS is that Sumé should become something of a 'model' for other irrigation schemes to follow.[32]

São Gonçalo was the second project studied and, as already stated, was one of the first generation of irrigation schemes set up by DNOCS (then IFOCS) during the 1930s. It lies within the *município* of Souza, Paraíba in the area of *sertão* known as the Depressão do Alto Piranhas (Higher Piranhas Depression), named after the river Piranhas which runs through the area and the project. Average yearly rainfall is quite high at 900 mm and is concentrated during the winter or rainy season from January to April. Even during the 1970 drought the valley received 500 mm of rainfall.[33] In terms of climate the area is privileged compared with some parts of the *sertão* where mean annual precipitation can be as low as 250 mm in a normal year. The Açude São Gonçalo was built in 1936 and, together with the more recently constructed Engenheiro Ávidos reservoir, has a capacity of 300 million cubic metres. As well as supplying the project, they serve the nearby towns of Souza and Cajazeiras.[34]

Under the old system most of the benefits of irrigation accrued to a few large landowners. They purchased water from DNOCS very cheaply but were not involved in any of the capital expenditure entailed in reservoir- or canal-building. Of 4,700 hectares covered by the scheme, divided into 182 properties, 84% of the land belonged to private landowners who either cultivated it directly or rented it out to farmers. Three properties occupied 31% of the total area, while large and medium-sized holdings absorbed 40% of the total income (in cash and kind). DNOCS's direct participation in agricultural activity was limited to a small area of land let to tenant farmers and a research institute which conducted crop experiments.

Following the 1970 drought São Gonçalo, along with other DNOCS projects, received new impetus from the recognition of the failure of attempts until then to alleviate the so-called 'drought problem'. A feasibility study was commissioned which recommended expropriation of all private property within the project boundaries, in view of the incompatibility of private ownership with a rationalized (i.e. reformed) irrigation scheme meant to transfer benefits to a larger population.[35] The project now occupies 4,100 hectares, of which 2,100 will eventually be irrigated. By February 1975 600 hectares were in operation. A total of 345 *colonos* and their families was envisaged by 1976, the first of whom were installed in 1971, reaching the figure of 126 families by early 1975. The system of distributing land on São Gonçalo is basically similar to that on all DNOCS projects. Each family is given around five hectares (depending on factors such as cultivation plan and soil fertility), three of which are irrigated in the form of a family plot (*lote familiar*).[36] Two hectares consist of grazing land, held in the form of common pasture by the *irrigantes*.

São Gonçalo was chosen as a case study precisely because it formed one of the earlier type of irrigation project. As such, it was possible to compare the effectiveness of present-day irrigation strategy based on public ownership and direction, with the former type which did not alter the pre-existing social and economic structure of the valley based on private property. In addition, a wealth of information concerning the situation in the valley before 1971 was made available by the reports of the consultants (Hidroservice).

Morada Nova, in the state of Ceará, was the third project studied. It is situated in the valley of the river Banabuiú, the major affluent of the river Jaguaribe. The valley is typical of many in the interior; relatively humid and protected compared with the surrounding *caatinga*, it supports a dense rural population. Rainfall averages 700 mm per annum. A research team originally surveyed the area in 1962–5 with a view to using the Jaguaribe valley's 60 billion cubic metres of stored water for irrigation.[37] DNOCS has plans for irrigating large areas of the Jaguaribe valley as one of its 'priority regions'. It covers 75,000 square kilometres and has a population of two million. The agency plans to establish eight projects in the valley, eventually covering a total area of over 100,000 hectares and absorbing some 15,000 families.[38]

Morada Nova was the first of these projects and started operating in 1971. Of the 12,500 hectares that will eventually be controlled by DNOCS in the Banabuiú valley, 8,000 will be taken up by this scheme. One-third of the area will be divided up into irrigated plots while the remainder is to be used for common pasture. By September 1975 one-half of the irrigated land had been prepared and 253 of the eventual 550 families accommodated. The goals of 2,300 hectares irrigated and 550 families settled on the land was officially supposed to be met by the end of 1975 but this was clearly not possible. At least another two or three years' work will be needed to achieve this.

This particular project was chosen, again, because it has made considerable progress compared with the majority of DNOCS schemes, which were still in the planning stage. Only twelve of DNOCS's *perímetros irrigados* had started to function by the end of 1975. The agency regards Morada Nova as its showpiece par excellence and, as with Sumé, DNOCS presents it as an example of what it hopes to achieve with irrigation. Considerable human and financial resources have been devoted towards making Morada Nova work. If it is to set the pattern for future projects, knowledge of the way in which it has developed could provide lessons about probable developments later on. Taken together, my three case studies accounted for 20% of the total area irrigated by DNOCS and 45% of the families installed by the end of 1975.

4. Employment and stabilization of the rural population

When drought conditions begin to disrupt the agricultural calendar for thousands of semi-subsistence farmers in the *sertão* and the production of staple foods becomes impossible, one of the first visible effects on the population is the abandonment of the land and the search for alternative means of support. In 1970 almost half a million refugees from the drought were employed on emergency work fronts building roads in the interior, the usual method which the government uses in an attempt to provide temporary employment and stem the outward flow of labour to the towns. The creation of permanent jobs in the rural sector, 'fixar o homem à terra' as DNOCS calls it, is, therefore, perhaps the most important of the three goals behind irrigation in the North-East. However, the inescapable conclusion reached, even assuming that official employment targets are met, is that irrigation in the *sertão* will do little to create more permanent jobs for the impoverished rural population. Moreover, there is every indication that current irrigation strategy displaces far more labour than it absorbs, thus exacerbating rather than ameliorating the problem of rural unemployment.

Although there is an abundance of labour in the interior of the North-East, irrigation projects are capital-intensive. It takes two hectares of irrigated land to create one permanent job. This is equivalent to a level of capital investment per unit of labour absorbed of US$4,800 or over seven times the Brazilian average for agriculture.[1] The fact that schemes are capital-intensive means that they absorb little labour in relation to the capital expenditure involved in their implementation. Even if DNOCS's optimistic target of 22,000 families settled by 1980 is realized, this will represent only 1.3% of the calculated excess rural population in the North-East by that time (1,700,000). If the *sertão* only is considered, this goal will still be equivalent to a mere 2.3% of the excess low-income rural population by the end of the decade (952,000).[2] Assuming that plans for incorporating 274,000 hectares of drier *caatinga* (*área sêca*) are fulfilled in addition to the 100,000 hectares of irrigated land, the total impact will be minute in relation to the problem. The significance of the plans diminishes further still when one realizes that by the end of 1975 only 4% of the 1980 target had been achieved in terms of labour absorption.

Despite the small number of permanent jobs created by DNOCS irrigation schemes, the level of investment is formidable. Over the period

1975–9 alone the agency plans to spend three billion *cruzeiros* or US$375 million.[3] Investment into irrigation under the GEIDA proposals for the North-East is put at US$550 million at 1971 values.[4] This sum represents almost half the total capital formation in all sectors in the region in one year, although spending on irrigation will cover a decade.[5] With the delays typical of project implementation so far, though, capital costs are likely to be driven up further still while the number of jobs to be created remains constant.

A certain amount of labour is temporarily employed during the construction of irrigation schemes, mainly in the building of canals, roads and buildings. During 1970, in fact, SUDENE established a number of work fronts on such projects in cooperation with DNOCS. While the infrastructure of a project is being laid out there is no doubt that additional employment is created which helps to compensate in part for the jobs lost through expropriations. However, such work is only temporary and does little to stabilize the population permanently. Colonists also hire outside wage-labourers to supplement the manpower provided by their family labour. It is difficult to estimate the precise extent to which outside labourers are used, especially as no records are kept and there is a good deal of overlap, for instance one *diarista* may work for several families. However, the cost of hiring extra workers is quite high for the farmer's limited income to bear. Many farmers on the projects visited experience considerable difficulty in paying *diaristas*; one study has put the proportion of colonists who have such problems at 35–50%.[6] Consequently, there is reason to believe that the amount of temporary employment created is not high.

Although the overall impact of irrigation in creating stable jobs is low, the number of people forced into a state of unemployment by the process of displacement from their homes is considerably higher. On all the projects studied, the number of people obliged to leave the land they had worked because of expropriation was six times the number who have so far been reabsorbed by the irrigation schemes subsequently built in those areas. Such severe social costs are certainly not taken into account by the traditional methods of cost-benefit analysis used in the GEIDA programme calculations.[7] This situation is frequently a source of acute embarrassment to DNOCS personnel, and the displacement policy is difficult to justify if irrigation is genuinely intended to be labour-absorbing. This is reflected in the fact that officials were often unwilling to reveal the true extent of population movements and, on several occasions, provided information which, subsequently cross-checked, proved to be untrue. Feasibility studies usually avoid mentioning or grossly underestimate on paper the size of the forced exodus from river valleys which necessarily accompanies irrigation in its present form. Generally speaking it is regarded as something of a necessary evil. However, 'fixar o homem à terra' as a catchword becomes meaningless to the point of absurdity when it is realized that in order to accommodate 100 families DNOCS has had to expel 500 to 600 other families from their homes.

By their very nature, river valleys in the *sertão* are privileged areas which offer shelter from the extended dry season or drought, even if the rivers themselves are not perennial. Production by means of *vazante* agriculture enables basic subsistence foods to be grown under any conditions but the most severe, and during the drier part of the year when cultivation on the *caatinga* is not possible. As such, the rural population has become concentrated in the valleys. Furthermore, generations of inheritance and land sales have caused extensive fragmentation of properties. Large numbers of smallholders using their family labour, sharecroppers, tenant farmers and squatters are thus able to make an adequate if not substantial income by intensively cultivating their small plots. The effects of expropriation on this type of population in the three valleys studied will now be considered.

In São Gonçalo, the population of the hydraulic basin before the project reform took place in 1971 was approximately 7,000, consisting mainly of rural producers of various types and their families.[8] Expropriation and the forced emigration that accompanied it proceeded rapidly after 1971 so that today 123 colonists and their families occupy the farmland, constituting about 1,100 people. The project will eventually resettle 345 families, or 3,000 people, which will still represent less than half the original population of the area. Although there are other employees or *funcionários* on the project which makes the total amount of labour absorbed higher, as regards the population which gains its living from agriculture, DNOCS appears to be expelling more than it absorbs. This is especially true in the light of the fact that on São Gonçalo the target of families to be resettled on the project has fallen from the original 600 to the current goal of 345.[9]

Some displaced labour was employed by engineering companies on project construction but most were obliged to leave the immediate area in order to seek new homes and jobs. A large number are known to have migrated from the area completely because of the difficulty of finding local employment. However, the eventual destinations of these emigrants are largely unknown and it can only be surmised that they went to other rural areas or to the towns. Some stayed in the local region, their presence made evident by newly erected urban slums in the nearby town of Souza as well as in small communities along the side of the highway such as Marizópolis. The cost of living for such emigrants is substantially higher in town since they are no longer able to produce their own subsistence crops and must pay market prices for food. Sanitary conditions in the new *favelas* are visibly worse, with open sewers running down the centre of the mud streets.

In Sumé the rate of displacement appears to have been of a similar magnitude. Before the private land was expropriated, according to DNOCS estimates, seventy-three families inhabited the valley, of which forty-seven were smallholders.[10] This figure may well underestimate the original population of the area because in every other valley studied, non-landowning families outnumber property owners by two to one. No independent source of information was available in order to check this fact. However, assuming this estimate to be realistic, the displacement level is still considerable. The project has twenty-six families at present and will increase its total to thirty-five when fully operational. Similarly, in Morada Nova, by combining various sources of information it was possible to estimate the previous population of the valley at 3,000 families or 18,000 people.[11] This compares with a current absorption rate of 253 families, rising to 555 eventually.[12] A sample survey of 348 families made just before expropriation started in the Banabuiú valley revealed that only eighty-nine had made any plans for leaving the area and only 40% had any idea of what jobs they would take up afterwards.[13]

The system for compensating the dispossessed does little to rectify the imbalance caused by a heavy rate of forced migration from the hydraulic basins. Large numbers of families own no land in the valleys where they live and work, and are entitled to no payments of any kind to compensate for their loss of a house or a job. In the Banabuiú valley (Morada Nova) two-thirds of those expelled from the land were non-property owners, or some 2,000 families. This was true of São Gonçalo and, to a lesser extent, of Sumé also where one-third of families owned no land. Information obtained concerning two other DNOCS irrigation projects in Ceará, Curú and Ayres de Souza, shows that here also about 60–70% of expelled families had no right to any form of compensation.[14]

Property owners receive varying amounts of compensation per hectare of land expropriated depending on both the quality of the soils and on the amount of possessions on the land such as buildings, fences, cattle, and so on (*benfeitorias*). A table of prices was established by SUDENE in 1968 but this is periodically up-dated by DNOCS. The price of land is set according to a multi-point scale, which rises to a maximum of around Cr$260 per hectare of bare land for top quality alluvial soils (1975). Although it is generally recognized that the price of land in the *sertão* is low, it has been suggested in some quarters that the rates paid by DNOCS in these relatively fertile valleys are well below market values in certain cases.

The tremendous variation in the rates of compensation paid by DNOCS derives from the fact that, of the total amount paid out, 10% is for the land itself and 90% for *benfeitorias*, or improvements to the property. In Morada Nova the price per hectare varied from Cr$90 to Cr$8,000. In São Gonçalo landowners received twenty times as much money for possessions expropriated as for the bare land.[15] Most landowners affected by expropriation are smallholders with relatively few *benfeitorias*, but a minority of *fazendeiros* tended to monopolize the land and, by virtue of this fact, the lion's share of *benfeitorias*. In the Banabuiú valley 90% of the properties taken over were under five hectares. The smallest was 0.1 hectare and the largest 790 hectares.[16] In São Gonçalo three landowners possessed 56% of the area, while 67% of property owners owned less than five hectares.[17] Most of those who qualify for compensation therefore receive small amounts, the average being from a few hundred to one thousand *cruzeiros*.

As already stated, many of the properties involved have been fragmented through generations of inheritance and the absence of primogeniture. One result of this process is that in areas expropriated by DNOCS only 60% or less of landowners, on average, have legal proof of ownership. Without this documentation they can be forced off the land with no legal compensation even though they may be the *de facto* owners. These 40% or more have little hope of being paid, but they may take their cases to court if they have the resources to do so. This may take years, however, during which time the value of any compensation will have been seriously eroded through inflation. On all three projects visited, some smallholders had not been paid after as much as four or five years of waiting.

The net result of all this for the population obliged to move is, to say the least, not pleasant. Most get nothing in compensation for their involuntary exodus, while those who do qualify receive settlements which are hardly enough to enable a new start elsewhere on the same basis as previously. Only the elite of larger property owners may be said to receive enough money to compensate them for their loss, precisely those most likely to have other assets elsewhere to which they can turn.

A large proportion of dispossessed farmers and others working in the areas expropriated became ineligible to join the irrigation projects built there. In order to apply a farmer must be a head of a family, between the ages of nineteen and forty-nine and of Brazilian nationality. He should have made his living exclusively in agriculture and have insufficient land to make a decent living ('uma vida digna'). He must also have been living in the same place for at least three years and he and his family should be of proven 'good conduct'.[18]

Following Law 4,504 (1964) and Article 8 of Law 4,593 (1965), plots in expropriated areas must be distributed first to ex-landowners thus affected and second to agriculturalists who previously worked there. Apart from these groups who receive preference, only landless workers are officially allowed to acquire irrigated plots. Landowners from outside expropriated areas who have 'adequate' holdings, as well as public employees, are strictly prohibited from becoming members.[19] A way around this regulation is given to project directors, who are allowed to recommend applicants (up to 10% of the total) so that they do not have to go through the normal procedures. These may include the sons of irrigation farmers and 'those with suitable knowledge and capabilities for irrigation'.[20] This discretion was clearly exercised in Sumé, where 20% of project members consisted of farmers whose previous occupation had been principally outside of agriculture. Geographically speaking, recruits should officially be drawn firstly from the areas expropriated, secondly from neighbouring parts of the valley, thirdly from drier areas of *sertão*, and finally from 'other strategic areas'.[21]

As might be expected, the local atmosphere when expropriation is in progress is one of heated, sometimes violent opposition. A good deal of friction is caused between DNOCS and the nearby population which often persists for years afterwards. A case in point is that of Morada Nova, where expropriation was hurried and people forced to leave their land abruptly. DNOCS hired a private company (Codesplan), whose job it was to make sure that the farmers and their families left the area by the deadline which had been set, a matter of only a few months from start to finish. Some of those who were particularly reluctant to move saw their houses bull-dozed away in front of them. As the project director put it, 'O DNOCS pecou um pouco' (DNOCS sinned a little). Many of those affected left the valley in disgust, which is reflected in the low proportion of total applications to join the project from those forced out (28%), despite the fact that they receive preference officially. Of the valley populations forced to leave their lands because of DNOCS expropriations, only 4–5% overall subsequently join the irrigation schemes established there.

An interesting reflection of the way in which DNOCS alienated the local population in Morada Nova was provided by my visit to the small village of Pedras which now lies just outside the project boundaries. Because I was obviously an outsider and had arrived in an official car the inhabitants thought, as I subsequently learned, that I was from DNOCS and had come to inspect the area in connection with a feared expropriation. Only after I had made my independent status clear did the villagers feel free to offer their opinions on the impact of the project.

One lesson that the agency learned from their experience in the Banabuiú valley is that much greater care must be taken over the way in which peoples' lives are affected by expropriation. Plans exist, for example, to extend the Morada Nova project up the valley of the Lower Jaguaribe. Its past experiences and opposition from the local clergy appear to be factors which are causing DNOCS to re-think its plans. Population density in the Jaguaribe is even higher than in the Banabuiú valley and the potential level of displacement much greater if DNOCS goes ahead with its intention of colonizing the valley on a similar basis.

The harsh and arbitrary nature of this expropriation policy is exacerbated by the fact that DNOCS has no plans for relocating the dispossessed. Although the need for such a *política de occupação* has been recognized by some of the agency's staff, DNOCS is run by civil engineers who appear to have little perception or concern for the plight of those forced off their lands. The agency would do well to take a leaf from CHESF's book in the São Francisco valley, which has made such contingency plans. When the enormous Sobradinho dam is built upstream from the twin cities of Petrolina and Juazeiro, former inhabitants of the flooded area will have the option of moving to new towns such as Remanso or to *agrovilas* such as Bom Jesus de Lapa.

In São Gonçalo the movement of many farmers and their families from the project area increased the pressure on land outside. The influx of *retirantes* during the 1970 drought intensified this pressure still further. In Morada Nova a similar situation occurred. Not only was the valley crowded with those seeking shelter from the drought, but the government established a work front here which employed 15,000 men, providing an additional source of attraction. The pressure on resources was such that after the first harvest of beans on the project there was a near riot by the hungry *flagelados* who threatened to raid the stores. Only the intervention of the project director and the distribution of food averted a potentially disastrous situation.

Another side-effect of the expropriation policy is that banks are now refusing loans to landowners whose properties are at all in danger of being taken over by DNOCS. If lands were expropriated with little warning, as happened in Morada Nova, they would be likely to lose their money and be unable to recoup their losses without considerable delays.

Discussion so far has focussed firstly on the small number of permanent jobs created by DNOCS irrigation schemes, and secondly on the disproportionately high rate of displacement from expropriated areas of farmland. An additional consideration is whether or not those few families absorbed by the projects were in fact vulnerable to the drought before joining. It would seem to be a logical corollary of irrigation in the *sertão* that if the strategy is to have any impact on the drought, however marginal, it must reach those sectors of the rural population which are liable to be adversely affected by abnormally dry periods. The evidence gathered on three case studies suggests overwhelmingly that this is not the case. By examining the geographical and social origins of recruits to irrigation schemes a common pattern repeated itself in all three areas studies: (1) very few *colonos* had ever been obliged to work on emergency fronts or migrate in order to seek employment during a drought, (2) most *colonos* came from within their respective valleys and were thus relatively protected from the drought, and (3) the drought victims are still untouched by irrigation directly, although projects may provide some shelter for a few *flagelados* temporarily.

The information concerning recruits' backgrounds was collected by means of both personal interviews with *colonos* and analysis of the interview questionnaires completed when applications are made to join a project. All applicants to irrigation schemes are interviewed several times, providing a fairly comprehensive picture of their economic backgrounds, family size, age and recent movements. By combining these sources 100% samples were obtained on all three projects studied. On the smallest scheme, Sumé, it was possible to interview all twenty-six *colonos* personally, but on São Gonçalo and Morada Nova the personal dossiers of all project members were made available and consulted to provide most of the necessary information. Each applicant may have as many as four interviews, so that some cross-checking was possible to test the consistency of the answers. Although at times the information contained in the questionnaires appeared to be a little confused regarding the precise nature of an interviewee's type of tenancy arrangement, on the whole the data were accurate and comprehensive. Information from individual projects will now be considered.

On the Sumé project, situated in one of the driest areas of the *sertão*, only four of the twenty-six colonists had ever worked on an emergency front and only two had ever left the area completely to find employment. The remainder had pursued occupations unaffected by drought or lived in areas relatively untouched by its effects. These included eight DNOCS *rendeiros*,[22] several inhabitants of the valley and one colonist who had previously worked on an irrigation scheme in the São Francisco valley. Other project farmers embraced an amazing variety of previous occupations. Among them was a driver who had worked in DNOCS for five years before joining Sumé, a barber, a market stall owner and a bar owner.

It is only fair to add, however, that Sumé was not typical of DNOCS projects in this respect. The agency's regulations state quite categorically that recruits to projects should have devoted themselves exclusively to agriculture, which clearly the latter group had not. Patronage may have influenced recruitment, or it could be that other local farmers were antagonistic towards the project because of the expropriations and population displacements and were reluctant to apply. Despite the fact that farmers from the expropriated area are officially supposed to receive preference in the selection procedure, only nine of the twenty-six fell into this category. One of the major reasons for this, apart from the disaffection of the local population, is that many of the dispossessed are ineligible to join and fail to meet DNOCS entrance requirements by virtue of their age, level of agricultural experience (sic) and even 'personality'. This procedure is discussed more fully in chapter 6.

In São Gonçalo, 84% of the 123 recruits to the project came from within the hydraulic basin which had benefitted from the old irrigation scheme before 1971. These included a variety of agricultural occupations: DNOCS *rendeiros* or *vazanteiros* (26%), DNOCS *diaristas* (15%), tenant farmers (4%) and ex-construction workers (5%), all of whom were protected from the drought effects. Sharecroppers from the drier areas of *caatinga* outside the project boundaries (*faixa sêca*) form one of the groups most adversely affected by the drought, yet they are strongly underrepresented on the project. In São Gonçalo they accounted for 63% of applicants yet only 14% of recruits. These figures undoubtedly represent official DNOCS policy of giving preference to those immediately affected by expropriation. The familiarity of the population with irrigation also helps to explain the high percentage of recruits from the expropriated area. Smallholders accounted for 6% of applicants to the project but no recruits. The low proportion of *minifundista* applicants may be explained by the widespread use locally of sharecropping as a tenancy arrangement, as well as by the fact, again, that many are ineligible to join. Some 6% of colonists are, however, sons of expropriated landowners.

Of the total number of recruits to São Gonçalo, 81% were born in the immediate locality and the remainder had, without exception, been living in the area for periods of between two and over twenty years at the time of joining the scheme. The Hidroservice survey also confirmed that São Gonçalo has continued to attract migrants from drier areas of the *sertão* since the old irrigation scheme was started in the 1930s, and especially in drought years, such as 1942, 1958 and 1970.[23] On the old scheme 42% of heads of families were from outside the municipality of Souza, in which the project is situated. 80% of the population had lived there for over two years, and over 50% for more than ten years.[24] Of the 42% from outside the immediate locality, one-fifth had arrived during 1968–70, evidence that the *açude* and fertile valley continued to attract people from less favoured areas of the interior.

Quite apart from the fact that reform of the old irrigation system caused the net expulsion of over 5,000 people from the river basin, there are two additional considerations. Firstly, most recruits to São Gonçalo were not adversely affected by the drought when they joined the project. Secondly, the area has been offering relatively stable employment for over forty years to both the locally born population and to *retirantes* from other areas of the *sertão*. It therefore seems quite clear that the project after the reform of 1971 has made little additional contribution towards stabilizing the rural population.

A similar pattern of recruitment was found to exist in Morada Nova where only 3% of colonists had ever worked on an emergency front and 90% came from within the Banabuiú valley, in which the project is located. Like other river basins, the Banabuiú has always offered a degree of protection from the rigours of the extended dry season. Smaller producers had the advantage of growing certain resistant xerophytic crops such as the *carnaúba* palm, which is highly commercializable. It can be used to produce a variety of by-products from building materials to wax. Before 1970 *carnaúba* production in the valley was second only to cotton, 65% of total production being located on properties of under 50 hectares. Cattle and cotton production (70%) were concentrated on holdings of over 50 hectares, affording additional protection against the drought effects.[25] Of these recruits from within the valley, 56% were sharecroppers, 15% ex-landowners and the remainder other types of rural labourer.

The tendency to select farmers who are not badly affected by the drought is accentuated by two factors other than the obvious one, already mentioned, of preference being given to those from areas taken over by DNOCS. Firstly, when a farmer applies to join a project his character has to be vouched for by friends or acquaintances living in the area. If an applicant can receive the personal recommendation of an existing *colono* he stands a far better chance of success than a farmer with similar qualifications but no such backing. Consequently, a self-perpetuating process is set in motion in which recruits are drawn from a limited geographical area and from restricted circles or networks or friends.

A second factor which contributes towards this pattern of recruitment is that, in order to minimize its administrative costs, DNOCS prefers to select people from areas as near to the project as possible. The expense of setting up recruitment offices and paying for transport, and simply the time taken to locate farmers living in out of the way areas, is thus kept down. In Morada Nova in 1975, for example, selection was transferred away from more distant municipalities such as Quixeré and Tabuleiro do Norte, back to that of Morada Nova itself. Given this logic, rural producers from the more distant, drier areas of the *sertão* who are most vulnerable to the drought, are virtually untouched by irrigation schemes under the present system of selection. Many of the dispossessed farmers to not wish to apply to join the project, discouraged by DNOCS's heavy-handed methods used to clear the valley. Although 3,000 families were affected, only 407 applications to join the project had been received by the end of 1975 from amongst those forced out of the area. Nearly three times that number applied who had not been touched by expropriation, but the vast majority of whom still inhabited the valley.

The reputation of DNOCS for efficiency is not strengthened by the fact that long delays are common in absorbing families even after they have been accepted. This is partly due to hold-ups in the preparation of the land for occupation, caused by bad weather conditions during the winter months which make soils very muddy. Other reasons are bad planning and coordination between DNOCS and private contractors, as well as bureaucratic delays in the authorization of funds and making of decisions. Delays are, furthermore, caused by the many health tests that farmers must submit to before starting on the project. Irrigation schemes therefore experience quite a high drop-out rate after acceptance since an aspiring *colono* may often have to wait up to three years before being able to settle on his new irrigated plot.

Expropriated farmers who have reached a certain age when they lose their plots of land are liable to find themselves in an extremely difficult position. If a farmer is over forty-nine he does not qualify to join the irrigation scheme. Unless he finds alternative employment or is supported by his family, the only other means of support would be the state pension scheme under FUNRURAL.[26] This allows retired rural workers a pension equal to half the minimum salary. However, this may not be claimed until he has reached the age of sixty-five, leaving a gap of sixteen years or more in which the dispossessed farmer has to get by some other way. In addition, anyone claiming a retirement pension under FUNRURAL must present documentary proof of his age and be officially registered.[27] Consequently, many do not have the legal right to even this state support.

In conclusion, therefore, basic similarities in the settlement patterns of the three projects analysed enable certain general tendencies to be identified concerning the nature of DNOCS irrigation. Firstly, it is quite clear that those poorer farmers who are periodically forced to seek jobs on emergency work fronts or in the towns when a drought strikes, are virtually untouched by irrigation. The very limited hectarage put under irrigated cultivation so far has benefitted rural producers who already enjoyed relatively privileged positions, protected from the excesses of climate and rarely forced off their land by a drought.

The evidence from these projects very strongly suggests that valley farmers are more likely to be expelled from their land by DNOCS rather than by any climatic vicissitudes. It is, perhaps, just as well that DNOCS's rate of progress in irrigating the *sertão* has been rather slow, since this form of project appears to create more unemployment than jobs, displacing thousands and reabsorbing a fraction of that number. This, of course, stands in direct contrast to the government's intention that irrigation should help to stabilize the rural population and help stem the rural–urban migratory flow.[28] It is also in direct opposition to DNOCS's own stated goal of tying rural producers to the land, 'fixar o homem à terra'.[29] The agency is quick to publish self-congratulatory brochures boasting of the number of families absorbed by projects, but omits to mention at what cost these relatively small achievements are made.

The disruptive nature of DNOCS's intrusion into rural affairs is exacerbated by the fact that it has no policy for relocating the dispossessed or assisting them in any way whatsoever to re-establish themselves in other areas. Neither is there any sign for the moment that any official change of tactics is planned. The closest thing to a modification of policy has been suggested by an outside institution, which recommended the establishment of health programmes in two planned irrigation schemes in order to minimize the rate of displacement.[30] However, it is extremely unlikely that health and sanitation improvements would alone be enough to keep people in the area unless alternative sources of employment were found for the dispossessed.

Local political pressures and public opinion may well force DNOCS to re-think its approach. There is already some indication of resistance to DNOCS strategy in the Lower Jaguaribe valley, Ceará, where the agency plans its largest single concentration of irrigated land in the North-East. It is, nevertheless, heavily populated and the social costs involved combined with local opposition after the Morada Nova experience will probably mean that some modifications of current tactics will be inevitable. For example, DNOCS recently abandoned its plans to expropriate a very heavily populated area of land at the end of the Banabuiú valley, adjacent to the town of Limoeiro do Norte. If a change of policy is not forthcoming, the extension of irrigation could be held up for a long time, prejudicing the agency's chances of receiving a continued high level of funding from the government. For the present, however, the Ministry of the Interior firmly believes that irrigation can make a major contribution towards alleviating the problems of rural unemployment associated with drought in the North-East.[31]

5. Rural incomes

Apart from creating more rural employment, a second and parallel objective of government irrigation in North-East Brazil is that of substantially increasing family incomes. A basic tenet of the GEIDA proposals is that irrigation schemes will earn large incomes from the sale of high-value fruit and vegetables in regional, national and international markets. It was predicted that by 1976 these crops would be finding their way to Europe and the USA, reaching a value of almost US$50 million by 1980, when it was expected that all projects would be fully operational.[1] Such high returns, it is thought, would justify the substantial capital investment in irrigation, and produce favourable cost-benefit ratios. They would also allow the incomes of families working on irrigation schemes to be increased to four times the average for non-irrigated agriculture in the North-East. In 1980 this would amount to Cr$2,100 (US$360) per capita as opposed to the average of Cr$583 (US$100).[2]

Other official studies of irrigation make similar forecasts regarding the ability of projects to generate higher incomes. DNOCS confidently predicts that family incomes on irrigation schemes will be six times the official minimum salary.[3] Firms of private consultants contracted by the agency to undertake feasibility studies of valleys considered fit for irrigation also maintain that rural incomes will increase by leaps and bounds. Hidroservice concluded in São Gonçalo that if the government bore the costs of project recuperation (which in fact it does on all irrigation projects) colonists could expect an income five times the minimum salary.[4] A similar level of income was predicted by SCET for families on Morada Nova.[5] In another study, the same foresees a rise in farmers' incomes of two to six times existing levels.[6]

Higher earning capacities such as these would bring several additional advantages. Firstly, they would result in higher living standards for the rural population benefitting from irrigated agriculture. Secondly, they would discourage rural–urban migration, which is considered excessive.[7] In the words of one survey commissioned by DNOCS, 'Irrigation . . . will permit the maximum creation of new jobs with higher incomes than before and will allow stabilization (sic) of the rural population, eliminating the climatic factor and giving a better distribution of income . . . this will prevent excessive rural–urban migration which industry does not have the capacity to absorb.'[8] A third effect would be to stimulate economic activity locally as a result of irrigation farmers' increased purchasing power.

However, evidence concerning income-earning capacity and income distribution which was collected on the three case studies does not support these optimistic forecasts. Before explaining the situation in some detail, the method used by DNOCS for organizing production and income receipt on its irrigation projects should be described. When a new member enters a scheme he is confronted with a bare plot of land that has been prepared by engineers for cultivation. In order that he may begin work he is given credit by the project cooperative to finance costs of preparing the land for planting, seeds, fertilizers, pesticides and the cost of hiring farm machinery, which is held in a communal pool organized by the administration. In addition, the colonist is advanced money to pay for the cost of hiring extra-familial labour, as well as a certain amount to cover living expenses while he is waiting to receive his income from crop sales. All agricultural produce is officially marketed by the project cooperative immediately, or stored for future sale. The accounting department of the cooperative later deducts the amount loaned to cover costs and family maintenance from the sum netted by marketing the colonist's crops. This difference represents the farmer's net profit for the year (*saldo líquido*).

Taking net profits as an indication of the level of economic success of irrigation farmers, the general tendency is for most to be in debt even after a year or more of belonging to a project. This means that the revenue derived from crop sales has been insufficient to cover the costs of production and family support, so that most *colonos* operate in the red according to the official accounts drawn up by the cooperatives.

During 1974 on the São Gonçalo project only 17 of the 114 members showed a net profit. It should be added that fifty-seven colonists had joined the scheme during that year, so that they would not have had a full twelve months' income to set against production costs. Nevertheless, when the remaining sixty-six were considered (those who had spent between one and two years there), it emerged that 75% of this number were also in debt. The size of individual debts ranged from Cr$700 to Cr$14,000. Amongst those who showed a profit, furthermore, the distribution of income was very unequal, varying from a *saldo líquido* of Cr$600 to Cr$24,000. Of the seventeen who showed a profit, the average income was exactly Cr$9,000 and the median income Cr$6,790. The difference between the highest income and the heaviest debt was Cr$38,000.[9]

On the Morada Nova scheme the situation is similar. Of 224 colonists who had been on the project for over a year, only 35% (79) showed a profit at the end of 1974. The remaining 65% had debts ranging from

Cr$60 to Cr$8,000. Those farmers with a positive *saldo* enjoyed incomes varying from Cr$60 to Cr$9,000. The income range was therefore less extreme than in the case of São Gonçalo, with just under Cr$10,000 separating the best income and heaviest debt. For the profit-making farmers, both the average and median incomes were considerably lower than on São Gonçalo, at Cr$1,264 and Cr$810 respectively.[10]

Income on the third project, Sumé, did not follow so closely the trends discernible on the two larger irrigation schemes. It was clearly the most successful of the three projects financially, which explains to a large extent why DNOCS regards Sumé as a 'model'. In this sense, although indicative of what the agency would like to achieve on its irrigation projects, Sumé was not typical. As I visited the scheme in December 1974, figures regarding income were only available for January to November of that year. However, as 1974 drew to a close twenty-one of the twenty-six colonists (81%) were enjoying a net profit. Sumé, for special reasons, was the exception in having a majority of its members show a profit. These reasons will be discussed in greater detail in the following chapter, but basically it was because Sumé, unlike São Gonçalo and Morada Nova (and unlike any other DNOCS irrigation project), managed to successfully commercialize a high-value crop (tomatoes), which gave the scheme its high income-earning capacity. The average income of profit-making farmers on Sumé was Cr$10,230 and the median income Cr$8,100. Income distribution varied considerably, however, the size of *saldo* ranging from Cr$1,100 to Cr$19,300 and the size of debt from Cr$280 to Cr$9,000.[11]

Broadly speaking, therefore, a clear tendency is evident in the way income accrues to irrigation farmers. A small proportion of colonists enjoys relatively high incomes, a greater number have middle-range earnings, while the majority are in debt. Sumé, as stated above, is not typical of DNOCS projects since it is the only scheme whose farmers show such a high level of profit. São Gonçalo and Morada Nova, on the other hand, are more typical because their patterns of production and commercialization have more in common with the general situation on DNOCS projects.

In São Gonçalo just 9% of the sixty-six longer-standing project members accounted for 85% of the project's income in 1974, each having a *saldo líquido* of over Cr$10,000.[12] On Morada Nova, 5% of profit-making farmers, who made over Cr$5,000 each in 1974, accounted for 27% of the income.[13] In Sumé the top 23% of farmers with a net profit took home 44% of total income, while the bottom 23% received 7%.[14]

On irrigation projects, then, it was found that a small majority of perhaps 5% earns between two and six times the official minimum salary, in accordance with official targets. The remainder, however, receive much less. Of those who show a net profit most earn well under the legal minimum salary, while the majority of farmers are in debt. This vast difference in incomes is reflected in the life-styles of *colonos*. The more successful acquire modern consumer durable goods such as radios and furniture for their homes. A few even buy jeeps with the proceeds of their agricultural labours. One colonist in Sumé was the proud owner of a new TV set, refrigerator and sewing machine, and was the envy of many DNOCS employees. The most successful farmers earn a good deal more than the lower grade *técnicos* and office workers in the project administrations.

This is not true for the vast majority of irrigation farmers, however, who must operate in the red in order to feed their families, and who cannot afford basic necessities such as water-filters and gas for cooking. Overall on the three projects visited, 63% of farmers were in this position. The true proportion may well be larger than this since some of those who had spent less than a year on a project would undoubtedly have been joining the ranks of the indebted during their second and subsequent years. The reason for suggesting this is that a large part of the debt for any given year consists of a carry-over of unpaid debts from the previous year. In Morada Nova, for example, 61% of colonists' debts in 1974 had been carried forward from 1973.[15] Two of the projects examined here, Sumé and Morada Nova, were also studied during their initial phases by another researcher. She concluded that at this time between 40% and 60% of farmers were operating in the red.[16] My more recent evidence appears to indicate that this proportion may be at least constant if not actually growing as agricultural production expands.

The clearly precarious economic position of many colonists on DNOCS irrigation projects has several unplanned side-effects. One is that the farmers market a significant proportion of their produce privately instead of handing it over to the project cooperative, a practice which is strictly prohibited by the regulations. This has several advantages for the colonist in need of cash. He receives his money immediately since he sells his goods directly to the wholesaler or public and does not have to wait until the end of the year. In addition he does not have to pay taxes (*Imposto de Circulação de Mercadoria* at 16%), FUNRURAL contributions (2%) or internal administrative charges (2%). At the same time, the farmer may be obliged to sell his crops outside for a price below the market rate, and also runs the risk of being asked to leave the project. Clandestine sales are discouraged by the administration of DNOCS which, in Morada Nova, went to the lengths of erecting a guarded entrance to the project complete with barriers, reminiscent of a border post. However, the practice continues unabated.

Of course, the fact that a significant, if unknown, percentage of farmers' crops is marketed privately will tend to hide true income levels. The official accounts quoted above are liable to understate the real degree of profit for some and exaggerate the level of indebtedness for others. The obvious response that this phenomenon provokes is to ask whether clandestine sales are motivated by farmers' poverty or whether this practice produces the negative figures on the balance sheets in the first place. The question is awkward to answer because by its very nature such marketing is almost impossible to measure.

My reasons for believing that it is fairly limited in scale are as follows. Firstly, if a large part of total produce were syphoned off privately it would provoke much harsher official opposition. After all, poor marketing figures reflect badly on the *gerência* or administration. Secondly, farmers are forced to resort to other practices which testify to their low incomes generally, despite any little extra gained through private sales. Many only manage to survive by purchasing food and other provisions on credit from local shops, frequently at prices up to 35% above the going rate. Thus, on official irrigation projects we see reproduced a phenomenon which is common in non-irrigated agriculture amongst semi-subsistence farmers who do not have sufficient resources to get by on without going further into debt. Food is also given by relatives to irrigation farmers with no ready cash to buy their own.[17]

Another method that colonists use to reduce cash outlay is to pay their hired labour on a sharecropping basis. Instead of paying a *diarista* the regular money wage a colonist may remunerate him partly in cash and partly with a share of the produce. The cost of hiring labour is one of the heaviest expenses that irrigation farmers have to meet, especially for those who have little help from their families. Day-labourers are often willing to work under this arrangement partly because of the difficulty of finding work outside and also because the experience thus gained in irrigated farming will stand them in good stead if they should decide to apply to join the project themselves. One study has suggested that, because of this, day-labourers receive less in real terms than they would normally.[18]

One of the main criteria DNOCS adopts to evaluate the 'success' of its irrigation projects is financial profitability, expressed by feasibility studies in the form of favourable cost-benefit returns. As we have seen, economic success judged in terms of increased family incomes has been rather limited so far. A warning that cost-benefit ratios had been unjustifiably optimistic as calculated in the GEIDA study (*PPI*) was given several years ago, however, by an independent analysis. Using the same GEIDA statistics, Cline estimated that of forty-seven projects selected for examination only seventeen had a favourable cost-benefit relationship.[19]

Nevertheless, the government is anxious that projects should be seen to be profitable, and helps colonists to minimize production costs in several ways. The Bank of Brazil (BB) and the Bank of North-East Brazil (BNB) charge negative interest rates (7%) on loans made to project cooperatives to finance production costs. Furthermore, repayments are not indexed to inflation (*correção monetaria*), unlike normal commercial loans.

The major cost item, capital expenditure on building canals, roads, and on levelling and preparing the land, does not appear on balance sheets in project accounts at all. Such works account for 57% of total investment into irrigation in the North-East.[20] Capital expenditure on irrigation projects in this region is quite high in relation to schemes in other parts of the world. It has been calculated, for example, that in the North-East it costs US$34,000 on average to settle one family on DNOCS projects.[21] Another study, based on the GEIDA estimates, put the cost per hectare at US$2,800 in 1972.[22] This compares with a figure of US$500 per hectare for irrigation in Europe, Morocco and Australia.[23] Cline suggests that the last figure does not include some costs which are contained in the Brazilian statistics. Adjusted accordingly, to cover costs of irrigation works and levelling only (excluding items such as farm machinery and housing), the level of capital expenditure per hectare in the North-East is US$1,450.[24] However, this is still almost three times the figure for Europe and Asia and is excessive by international standards. Irrigation in the North-East is basically capital-intensive and absorbs little rural labour in relation to the investment involved.

The costs of preparing projects have also increased rapidly because of delays caused by engineering problems, bureaucratic hold-ups and slowness in the selection of colonists. By the end of 1975 São Gonçalo should have absorbed its full quota of 350 families yet the number remained almost stationary throughout the year at 123.[25] Similarly, on Morada Nova, of one hundred additional farmers to have been recruited in 1975, none in fact joined the project.[26] Delays such as these in the São Francisco valley caused the cost per hectare to rise from US$1,450–1,950 in 1965 to US$2,400 in 1971.[27] By 1971 the Bebedouro project was costing US$2 million each year to maintain yet no farmers had been installed.[28] This was equivalent to over half the total amount budgeted for irrigation in the North-East that year.[29]

Another 'hidden cost' of irrigation schemes is the enormous bureaucracy which is supported entirely by the state. It is possible to argue that in a region such as the North-East, where there is little industrialization and employment opportunities are scarce, the government is justified in maintaining a large body of employees. However, the first priority of irrigation schemes is to benefit poor rural producers. It seems rather absurd, therefore, that in São Gonçalo, for example, DNOCS should employ well over 300 staff, yet the project has only absorbed 123 colonists. In Morada Nova DNOCS employs around 180 staff to serve 253 irrigation farmers. It is planned that this number will rise to about 230 when the project is fully operational and has 500 families.

Such investment and administration costs are borne entirely by the government. Originally, it was intended that, after three or four years, irrigation farmers should share this burden by purchasing their land and houses over a period of twenty to twenty-five years. Although, six years later, this matter is still officially under consideration, it is quite clear that the majority of colonists simply cannot afford such payments. Judging from the income figures it seems highly unlikely that they could withstand any more deductions beyond those made for running costs of production. This fact was confirmed by the administration privately, in conversation with officials from the accounts departments on the three projects, although it is not a statement of official policy. Even if the costs of land and housing were recouped, they would account for perhaps 30% of total capital expenditure.

There is a drop-out rate on DNOCS irrigation projects of about 5–10% of recruits who actually receive their plots and start producing. Those who leave are inevitably among the least successful economically and incur substantial debts. Indeed, when it appears that it is beyond a farmer's capacity to repay what he owes, he is invited to vacate his plot for another recruit. The fact that DNOCS simply cancels out this debt without its appearing on the balance sheet points to another significant 'hidden cost' of irrigation. This might appear, at first sight, to be an act of charity on the part of DNOCS. While this sentiment may not, indeed, be totally absent, the principal motivation for writing off bad debts lies elsewhere. It is to be found in the agency's concern, I suggest, for maximizing profits on paper, that is, in the official accounts. It is unlikely that heavy debts would, anyway, be repaid by an out-going colonist since repayment would be completely beyond his means.

In order to maximize paper profits, therefore, DNOCS has chosen to subsidize production costs in the ways described above and to strive for profitable harvests. It has done this rather than concentrating on minimizing investment and production costs. Nevertheless, the fact must be faced that most colonists operate in the red rather than in the black, and that the economic success of projects is questionable, even within the narrow cost-benefit framework of reference adopted by GEIDA and DNOCS.

The incomes of colonists as measured by their net profits vary considerably, as we have seen. It is difficult to say exactly why the income of a particular farmer is different from that of his neighbour who uses similar methods of production. Nevertheless, a number of major factors may be identified which have an important bearing on the distribution of income.

Perhaps the first major reason which helps to explain variations in the economic performances of irrigation farmers is differences in the amount of capital possessed by new recruits. Ex-landowners and their sons, for example, who are likely to receive some compensation for the loss of their property, will probably have more resources at their disposal than landless labourers who join. In Sumé ex-landowners and their sons accounted for 60% of project members, while in São Gonçalo the figure was 5%. In Morada Nova 15% of colonists were ex-landowners. Apart from this group, other recruits may enter with resources that give them additional support during the initial stages of production. In Sumé, for example, over 20% of farmers had been working in non-agricultural occupations before joining. It is difficult to determine the importance of this factor because it is correlated with others such as time of entry into the scheme and quality of the land. That is, ex-landowners and other 'better' recruits are liable to be among the first to join and to receive the more fertile plots.

A second factor which influences income differentials is variations in soil quality, particularly fertility affected by salinization. The sites of irrigation schemes were not chosen according to the criteria of highest soil fertility and consistency. It was, rather, the presence of a reservoir with the capacity to irrigate a given hectarage of land that was the main factor behind the choice of locations. It also meant that a large part of the capital expenditure had already been made. However, despite their relative fertility, alluvial soils in the *sertão* are highly varied in composition both horizontally and vertically. This creates a lack of consistency which has serious repercussions on controlled, irrigated crop production unless the planning and technical abilities of those involved are exceptional.

Salinization of soils is a serious obstacle to improving productivity. The soils of the *sertão* have a naturally high concentration of saline elements derived from the region's crystalline rock structure. Salinization takes the form of an impervious layer of saline, crust-line deposits which sits between 50 and 150 cm below the surface, rising sometimes to appear on top as a white substance. This layer must be broken up before the land can be cultivated, an expensive and time-consuming operation which, however, never solves the problem completely. Once recovered, saline soils are often not economically viable and are usually reserved for salt-resistant fodder crops such as elephant grass.[30] With good drainage it is possible to keep saline deposits well below the surface so that crops are not harmed, but if drainage of the land is inadequate, the water table rises and the saline moisture reaches the roots of the plants, severely reducing yields.

Salinization of soils, according to one estimate, affects no less than 30% of the land in the *sertão*'s valleys.[31] Large sections, totalling several hundred hectares, had to be taken out of production on all three projects visited. As a result, colonists who had incurred heavy losses through crop failure caused by salinization were transferred to other irrigated plots. Their debts, however, remained with them. One study concluded that the poor condition of plots when they were handed over to colonists explains, to a large extent, the poverty of the poorest minority of farmers.[32] I also found that those colonists with the heaviest debts had, almost without exception, experienced problems of poor drainage and salinization on their lands.

In São Gonçalo soil fertility is so varied that Hidroservice advised against a division into individual holdings because the firm felt that unlucky colonists with poor soils would be under a serious handicap. It suggested, instead, cultivation of the area as one single property as the best way of evening out the otherwise inequitable distribution of plots.[33] The company felt so strongly on this point that, even when asked to reconsider some of its conclusions by DNOCS after publication of the major study, it still stressed the advisability of a communal land-holding system rather than individual family plots.[34] It was added that distribution of different-sized plots as compensation for varying soil fertility would be of little use since during the preparation stage poor soils would require more work and expense, making it difficult to convince farmers with heavier cultivation and repayment responsibilities that their plots would be as productive as more fertile areas. The company also made the point that the tradition already existing in São Gonçalo for rural labouring under irrigated production obviated the necessity of having individual family farms. The project could, rather, be run by a cooperative and employ workers on a salaried basis.[35]

Despite the enthusiasm of the consultants for collective cultivation of the land, however, DNOCS ignored their advice and went ahead along established lines, distributing 5-hectare plots to individual families. The first batches of colonists on São Gonçalo received plots with low indices of salinization and nearly all showed a profit in 1974. Later recruits were given land which had heavier, sandier soils and greater problems of salinization. The level of indebtedness is much higher in this latter group, due mainly it seems to higher cultivation costs and reduced yields.[36] Until

February 1975 three colonists had been transferred from their original plots to other areas in São Gonçalo because poor soils had made cultivation impossible. Furthermore, 800 hectares of land were declared unrecoverable because of prohibitive costs, approximately Cr$30,000 to Cr$40,000 per hectare.[37] In both Sumé and Morada Nova areas were taken out of production and farmers transferred for the same reason.

Lack of technical ability to recuperate soils fast enough and well enough to ensure high crop yields compounds the natural problems of valleys whose suitability for irrigation is already dubious. DNOCS's problems in this respect have often not been helped by feasibility studies which have sometimes underestimated the variations in soil quality and, consequently, the costs of recuperation. The pedalogical study made by SCET in Morada Nova, for example, was based on an analysis of only the top 20 cm of soil.[38] Soil-type variations proved to be much greater than the study indicated, and had implications for the subsequent adaptability of crops chosen to suit certain areas. A study of 150 cm in depth would have provided a more realistic basis for planning crop production.[39]

Low yields caused in part by poor soil fertility and salinization have undoubtedly contributed to the fact that production targets have fallen far short of expectations. It was estimated that in 1975 the total value of agricultural production on São Gonçalo would be Cr$10 million. However, production during 1974 only reached Cr$1,750,000 and the number of colonists remained constant from one year to the next.[40] There was no possibility of these optimistic targets being achieved. In Sumé during 1974 income fell short by 20% of planned levels.[41] On Morada Nova this gap was considerably larger since by July 1975 only 18% of its total predicted income for that year had been netted, although half of the year's production had been sold.[42]

A third and related factor which helps explain variations in income amongst irrigation farmers is differences in production costs. These vary considerably per hectare of land under cultivation and are influenced by a number of considerations such as soil fertility, techniques used and types of crop cultivated. The cost element for irrigation farmers is made up of a number of inputs including seeds, fertilizers, pesticides, tractor hire and employment of outside labourers. Soil-type will have an important bearing on the amounts of seed and chemicals needed to ensure adequate yields. The fact that soil quality varies sharply causes parallel differences in the amounts of these inputs needed.

The cost of hiring additional farm labourers is one which many colonists have difficulty in meeting. A recent study concluded that between 35% and 50% of irrigation farmers experienced problems in hiring *diaristas* to supplement their family labour.[43] In tacit recognition of this difficulty, DNOCS tries to select families with several children old enough to help their fathers in the fields, theoretically minimizing the amount of extra labour that must be hired. It has been suggested that in practice recruitment of outside labour by irrigation farmers is not correlated with family size nor with the farmer's level of economic prosperity.[44] Variations in labour cost, it is maintained, are more the result of poor training by DNOCS technicians who do not instruct farmers properly on the use of additional workers.[45]

The cost of employing wage-labourers is, nevertheless, likely to fall more heavily on the shoulders of families as their children grow up, leave the land and go to work in the towns. Many colonists I spoke to had older children receiving their education in the larger towns, some even at college or university. While this may testify to the 'elitist' nature of recruitment to irrigation schemes, it is indicative of the fact that fewer siblings are likely to want to continue in their parents' footsteps as agricultural producers. In Morada Nova, for example, only 30% of parents anticipated that their children would stay on the land.[46]

This figure would probably be higher if the economic rate of return on irrigated land offered sufficient incentive, but not as long as most families continue in debt. It also raises serious problems for DNOCS who originally intended that plots of land should be passed on from father to son to take advantage of the knowledge of irrigation techniques acquired by children through their assistance on the land. If replacement farmers from outside are needed the time, expense and uncertainty of training new colonists from scratch will add to their problems. On the other hand, irrigation would then be more likely to reach those affected by the drought. If a self-perpetuating 'elite' of irrigation families is created this would not be so.

The training and supervision of irrigation farmers provides the fourth factor which affects income differentials. After being accepted into an irrigation project the new colonist is given a training course of just two weeks, which consists of lectures and practical demonstrations. Unfortunately there is often a long delay, possibly a year or more, between this training and the time that the farmer takes up his plot. A person's training might be rendered almost useless by this time. In São Gonçalo the twenty colonists trained in January 1974 could not hope to receive their plots before September of that year.[47] Inefficiency on CODEVASF's Bebedouro project in the São Francisco valley concerning selection and training has also been noted. By March 1971, although the agency claimed to have settled thirty-nine families, it was discovered that none had been enrolled or trained, and classes had been postponed indefinitely. Due to be completed by the end of 1970, it had not even been half-finished by the end of 1971.[48]

Once installed, colonists are supposed to receive regular supervision and assistance from rural technicians (*práticos rurais*). Such contact is frequently discriminatory and some farmers receive far more help than others. Distelhorst found that the better-off farmers tended to receive more attention than the less wealthy, which helped to generate mutual distrust between colonists and *práticos*.[49] Despite formal training, it was discovered that only 20–40% of irrigation farmers had adequate knowledge of soil-types and specialized techniques.[50]

Bad planning can sometimes be seen to compound the debt problems of colonists in very obvious ways. In São Gonçalo during the first year of production, for example, plots were prepared out of season. This caused delays which prevented the planting and harvesting of a banana crop, the project's main money-earner. This led to much heavier debts than would normally have been expected and was a severe liability on the farmer. On the same scheme the administration decided to prohibit the farmers from growing their own subsistence crops, going totally against tradition. Apart from causing widespread discontent amongst the colonists, their reaction was to grow their beans and corn secretly on spare patches of land such as on the sides of earth drainage canals. Given the fact that most farmers were in considerable debt, it seems surprising that the administration should have taken such action, apparently assuming that the farmers had the cash to buy food on the open market, which is more expensive.[51] Colonists on the other two projects studied were encouraged to grow their own subsistence crops.

Another phenomenon which pointed to the difficult economic circumstances of many colonists' families on São Gonçalo was that social workers received complaints about the hard time they had under the system of advance loans for family maintenance and labour costs. A significant number protested quite openly that they simply could not make ends meet and that their families' health was suffering as a result. The inconsistencies of this system, which is used on all DNOCS projects, derives from the fact that money for paying outside labour and family support is advanced together in one lump sum at regular intervals. This amount is calculated according to the number of man-hours estimated by the administration to be needed by a colonist for his land, taking into account the type of crop, methods of cultivation and so on. A table of fixed payments is drawn up and rigidly adhered to.[52] If a farmer has few children to help him it is assumed that he will have more money left over to spend on outside labour and he will need less for family maintenance. If the colonist has many children to help him, the money that would otherwise have been spent on paying *diaristas* is left to support his family.

However, although fine in theory, in practice the system is full of anomalies. This method is inflexible and takes no account of actual family needs. If a farmer cannot be sure of his children's help, through illness or whatever reason, he is forced to hire extra labour and therefore left with insufficient money for family support. Furthermore, in order to help their parents, children are often obliged to forgo their education, although some progress has been made towards adjusting the agricultural and educational calendars to minimize this. DNOCS frequently underestimates the number of man-hours needed on plots because of poor soils which require additional treatment, but for which the official table makes no provision. In São Gonçalo during 1974 families received, on average, only half the official minimum salary of Cr$265 per month with which to feed their families. It seems understandable, therefore, that farmers should resort to selling their produce privately for an immediate income rather than market their crops through the cooperative, which means waiting for months with possibly nothing to show for it at the end.

One determinant of income differentials which has so far not been discussed is the individual ability of the farmer. Adaptability to irrigated agriculture is obviously crucial, since for most recruits farming on a project represents a strong contrast to what they were used to as semi-subsistence producers. Such an analysis is, however, beyond the scope of this study. My intention has been, rather, to indicate broader trends and factors which characterize irrigation schemes. It was for this reason that, in addition to general data, limited case studies of three projects were made as opposed to a more detailed analysis of a single scheme.

DNOCS irrigation schemes may have been largely unsuccessful in raising farmers' incomes, but it should be pointed out that the agency has conferred other benefits to those absorbed by projects. In conjunction with state and local authorities DNOCS has established schools for farmers' children and has improved health care. Project members are encouraged to be vaccinated against diseases such as smallpox and cholera, while medical attention has been made accessible locally for the handling of routine complaints. Problems do arise on projects situated some distance from the nearest town, since schemes do not have permanently staffed medical posts. Treatment is given by local doctors under the FUNRURAL scheme in cooperation with DNOCS. In such cases, as in São Gonçalo for example, transport is usually arranged for the colonists by the social service division.

Education is another aspect of rural life which has been improved by the setting up of primary schools on projects, staffed by state-employed teachers. One obstacle to the children's education, however, is that their labour is often needed by parents on the irrigated plots, resulting in low attendance rates. Some progress has been made towards combining scholastic and agricultural calendars so that they do not conflict. There is, nevertheless, the possibility that an inbuilt contradiction exists between providing education for colonists' children and the official aim of trying to persuade irrigation farmers' sons to take over the plots in order to ensure some continuity. The likelihood exists that, once children are educated and literate, they will wish to leave agriculture and seek higher status employment in the towns. DNOCS would then have to spend additional time and money training recruits unfamiliar with irrigation techniques. This, however, is only a hypothesis, since it is too early to judge longer-term trends in the educational field of project activities.

The foregoing discussion on incomes represents, for most colonists, the failure of projects to fulfil the promises made to them as aspiring irrigation farmers. The continued indebtedness of the majority has left many little better off than previously and, if anything, serves to reinforce the suspicions of farmers outside, originally fuelled by the earlier expropriations and displacements. The inability of irrigation schemes to commercialize their produce with the profit margins predicted for them is perhaps the main reason for their lack of economic success, and it is to this aspect that discussion now turns.

6. Agricultural production

The third main objective of government irrigation schemes in Brazil is to increase agricultural production. This, it is argued, will serve two purposes. Firstly, it will satisfy excess demand capacity in future years and, secondly, it will bring in high incomes, justifying the heavy capital costs of investment. This high income-earning potential is based on the assumption that most revenue will be derived from the production and sale of high-value fruits and vegetables which non-irrigated agriculture in the North-East cannot produce on any significant scale. The GEIDA proposals concluded that in the North-East production of traditional crops such as cotton, rice and beans would be comparatively small because the demand for such crops is already well catered for by non-irrigated agriculture, and irrigation projects would have little relative advantage.[1]

In order to determine the feasibility of producing large quantities of fruits and vegetables, the *PPI* study calculated the probable level of excess demand for twenty-five products in 1976 and 1980. This was done by projecting trends which had been observed during the previous years in general crop production, productivity and areas cultivated.[2] The growth in population, said to account for 70% of the expansion in demand for food products, was also projected by rural and urban sectors. Per capita incomes were estimated for 1976 and 1980 on the basis of their growth during the previous twenty years.[3] Probable demand was set against expected production levels to give a set of figures showing the likely excess demand which irrigation will be able to help meet.[4]

Substantial margins of excess demand are predicted by the *PPI* analysis for all the major urban centres of the North-East. Although existing levels of consumption of high-value fruits and vegetables are still relatively low, it was predicted that traditional agriculture would be incapable of meeting the demand predicted for 1976 and 1980, especially in relation to oranges, other fruits, potatoes and other vegetables.[5] At the same time, however, limited internal markets would impose demand constraints on certain crops such as grapes, onions and tomatoes.[6]

On the basis of these calculations, GEIDA envisages a large and extremely diversified output for irrigation projects in the North-East. By 1980 citrus fruits, it is predicted, will be the principal source of revenue with a production of 603,000 tons per annum, followed by mixed vegetables (383,000), tomatoes (302,000), cotton (254,000), potatoes (225,000) and mixed fruits (102,000). The remainder, in order of importance, includes melon, fodder crops, grapes, bananas, sunflowers, onions, peanuts, beans, wheat, pineapple, garlic, rice, corn and sorghum.[7] High-value, non-traditional crops will account for 84% of irrigated production in the North-East, compared with 48% in the Centre-South.[8] It was estimated that the fifty-six projects in the North-East would generate an income of Cr$1,040 million at 1969 values, or US$248 million,[9] quadrupling per capita incomes for irrigation farmers and their families.

In its 1971 study GEIDA calculated that by 1976 irrigation projects in Brazil would be exporting their produce to Europe and the USA, taking advantage of their theoretical ability to supply high-value crops of superior quality during inter-seasonal periods in northern hemisphere markets.[10] By 1976 exports of fruit and vegetables to the USA, Canada, the EEC and Scandinavian countries would reach over US$19 million.[11] Grapefruit produced under irrigation in Brazil, for example, would account for 100% of Switzerland's imports of the fruit, 16% of France's, 10% of Germany's, 17% of the Benelux countries' and 11% of the UK's.[12]

It was also proposed that a new system of commercialization be established which would eliminate some of the intermediary stages between producer and consumer typical of traditional agriculture. This would reduce marketing costs and ensure that irrigation farmers received a fair return. In chapter 2 it was shown, for example, how small farmers are placed in an inferior bargaining position by their lack of resources to store and transport their produce. Very few are organized into cooperatives. Wholesalers and retailers of various types take the lion's share of the price paid by the consumer, as much as 60–70% or more.[13]

This new commercialization network was to be organized around local marketing centres to which projects would send their produce for classification, storage and processing. From these centres (*Centros Micro-Regionais de Commercialização*) the crops would go directly to supermarkets and other retail outlets for the domestic market, as well as for export. They were to be administered by a centralized institution (unnamed), possibly in cooperation with SUDENE, which has similar plans for improving marketing procedures.[14] The administration in charge of these centres would be firstly responsible for coordinating the collecting of agricultural produce from irrigation projects. Secondly they would organize the internal activities of storing and processing, and so on. Thirdly, and very importantly, they would be responsible for organizing the sale of produce, transport to markets and planning ahead with market research.[15]

There is no doubt that irrigated agriculture in the North-East has resulted in greatly increased yields for high-value crops at the experimental stage, as well as for more traditional crops in the actual production phase (see table 16). Following GEIDA's guidelines, DNOCS has made considerable efforts towards producing higher-yielding varieties, on the explicit assumption that only these crops will bring in sufficient revenue to compensate for the high capital and cultivation costs compared with the non-irrigated sector.[16] However, despite the desirability of such a production pattern, DNOCS has been unable, by and large, to grow and commercialize successfully high-value crops on a significant scale. Instead, projects have marketed crops which are already widely cultivated in the interior such as cotton, rice, maize and bananas. There is little relative price advantage to be gained with these crops, and this is perhaps the main reason behind the generally low income-earning capacity of irrigation schemes.

The crops most widely cultivated under irrigation have not been the citrus fruits and vegetables which head the list of GEIDA priorities. Cotton, rice and bananas account for about 70% of the total hectarage on DNOCS projects.[17] Data from the two largest case studies confirm the general trend. In São Gonçalo rice accounted for 55% of total revenue and bananas for a further 35% in 1974. Along with cotton, it was planned that these three traditional crops should bring in 97% of the project's income during

TABLE 16. *Comparative crop yields (tons/hectare)*

		Non-irrigated	
Crops	*DNOCS*	N.E.	South
Cotton	1.8*	0.3†	0.9†
Bananas	55.0	NDA	NDA
Tomatoes (table)	40.0	6.1†	21.0†
Tomatoes (industrial)	35.0	6.1†	NDA
Onions	20.0	0.6–0.8	5.0– 8.0
Melons	9.0	NDA	4.0
Carrots	18.0	NDA	NDA
Maize	12.0	0.70†	1.62†
Rice	4.5	0.8–1.9	1.5– 2.3
Sweet potatoes	10.0	4.6–5.5	1.2–22.0

Sources: Evandro de Souza Lima, *A Irrigação no Nordeste*, SUDENE, Fortaleza, 1973, p. 11.
*DNOCS, *Irrigação*, Fortaleza, 1975.
†IPEA, *Agricultura Brasileira: Comportamento Passado, Situação Atual e Perspectivas de Crescimento*, Brasilia, 1973, p. 20.

1975.[18] In Morada Nova during 1975 cotton (47%), rice (19%) and maize (16%) earned 82% of the scheme's revenue.[19]

Sumé is the exception to this general rule in that it obtains 80% of its income from selling tomatoes, a crop which is not produced locally outside the project nor in the *sertão* as a whole. These are mainly industrial tomatoes sold to canning factories, but a smaller proportion is marketed for domestic consumption. The relatively high yields on Sumé combined with the facility of producing two harvests per year, gives the project an advantage over other non-irrigated areas in the *agreste* region which produce tomatoes. At the same time, irrigated production in Sumé does not compete with local agriculture, competition which might depress prices. Although not without its disadvantages, tomato production in Sumé has assured the project a relatively high income, very much along the lines envisaged in the *PPI* plans. It must be borne in mind, however, that the distribution of this income is highly concentrated into the hands of a very small minority, for reasons already outlined in the previous chapter.

It is not easy to say exactly why Sumé has had more commercial success than either São Gonçalo or Morada Nova, or indeed any other DNOCS project. I would suggest that the reason lies principally in the fact that considerable technical resources have been devoted towards making tomatoes grow successfully here. At the same time, the project has remained very small (twenty-six families) and easy for the administration to supervise and control. This has made it simpler to coordinate production and marketing, especially having found a few outlets for the crop at what the administration reckons to be good prices.[20]

A major explanation for the inability of DNOCS projects to commercialize high-value fruits and vegetables is the fact that feasibility studies appear to have overestimated the probable effective demand for such crops. Market limitations have undoubtedly been partly responsible for the changed emphasis away from luxury crops to more traditional varieties for which markets and commercialization channels already exist. If planned output of major luxury crops is set against projected demand capacities it seems clear that these expectations are unrealistic. Comparing columns B and E of table 17, for example, it may be seen that the planned output of melons is fifteen times the prospective demand in 1980. Irrigated production of tomatoes, pineapples and onions is equivalent to two-thirds of market capacity. Other studies have contradicted the GEIDA conclusions, suggesting that rather than a high level of excess demand for certain crops (see table 18) there will be a surplus production even without any further contribution to supply from irrigated agriculture. This applies to bananas, pineapples, grapes and melons.[21]

TABLE 17. *Prospective demand and planned output: North-Eastern irrigation projects, selected crops*

Crops	*A*	*B*	*C*	*D*	*E*
Grapes	10,000	228	19	539	920
Tomatoes	3,000	804	18	775	1,244
Pineapples	2,500	313	12	338	541
Melons	3,000	334	11	12	22
Onions	3,300	274	9	273	436
Potatoes	3,100	240	8	1,606	2,394
Rice	520	NDA	NDA	NDA	NDA
Cotton	48	NDA	NDA	NDA	NDA

Source: William R. Cline, *Irrigation Projects in Brazil's Northeast: Case Studies in Social Cost-Benefit Analysis*, IPEA/INPES, R.J., 1971, p. 27.
Notes:
A. Profit per hectare (Cr$ at 1969 values).
B. Planned production ('000 tons, 1980).
C. % of total production value.
D. Total production, Brazil, 1968 (000 tons).
E. Estimated demand, 1980 (000 tons).

TABLE 18. *North-East Brazil – projected excess demand in 1976 and 1980 (000 tons)*

Crops	*1976*	*1980*
Avocado pears	12.8	16.3
Pineapples	12.3	17.1
Bananas	48.0	55.0
Oranges	695.5	866.4
Limes	23.0	34.2
Tangerines	38.6	52.3
Grapes	34.2	54.5
Other fruits	111.2	115.5
Garlic	39.6	52.5
Potatoes	351.9	490.8
Onions	82.1	112.4
Tomatoes	145.3	222.3
Other vegetables	338.1	451.6

Source: GEIDA, *Programa Plurianual de Irrigação (PPI)*, vol. II, *Estudos de Mercado e Commercialização*, annex 3, São Paulo, 1971, p. 6.

These demand constraints are reflected in the current production pattern of irrigation projects, where high-value crops account for perhaps 10% of total income instead of the predicted 77%. Paradoxically, this changed emphasis may prove to be roughly in line with future needs if independent market studies are correct in their calculations. It has been forecast that by the end of the decade there will be a deficit in the supply of various traditional crops such as cotton, rice and maize.[22] From the point of view of economic returns, however, the profit-making potential of these crops will always be limited compared with luxury products.

At the level of individual projects there is also a considerable difference between existing production patterns and those suggested by respective feasibility studies. Although traditional crops account for 90% of total production in São Gonçalo the pilot study placed far greater emphasis on the marketing of higher-value fruits and vegetables such as *goiaba* and onions, which could be commercialized during the inter-harvest period outside.[23] Similarly in Morada Nova, large-scale production of fruits and vegetables was predicted on the assumption that 'Production in the Banabuiú valley will always be inferior to potential demand.'[24] In reality such crops form an insignificant part of the project's output. Part of the reason for this discrepancy must lie, once again, in the over-optimistic predictions of future demand. Indeed, it is not clear whether these estimates are based on calculations of effective demand which take into account consumption patterns, income elasticities and competition from other sources, or whether they simply represent normative guesses as to what the population should be able to absorb. The latter alternative would seem to be nearer the truth.

Having laid some of the blame for existing production patterns on the optimistic predictions of feasibility studies, both at regional and project levels, it should be added that this is only part of the answer. Whatever the precise forecasts of market potential, there is undoubtedly room for expanding the production and sale of high-value fruits and vegetables. To cite one example, São Paulo competes strongly with local suppliers in Recife, providing this North-Eastern city with 92% of its potatoes in 1970.[25] This crop could be produced under irrigation in the North-East instead of its coming over 2,000 kilometres by road from the south of the country. Another reason, therefore, for the failure of projects to sell more high-value crops lies in their lack of technical and organizational ability to produce and market these crops. This problem is compounded by the increased size of irrigation schemes as they expand and become more difficult to manage with the resources that DNOCS has at its disposal.

The limitations imposed by such factors on any possibility of diversifying production along the lines suggested by GEIDA become apparent when the record of attempts to market high-value crops is examined. In order to obtain the best possible prices and to maximize incomes it is necessary to market crops produced by irrigated agriculture during the off-season outside. Bad planning, however, has often resulted in these potential advantages being thrown away. During 1973 in São Gonçalo, for example, a 30-ton harvest of onions had to be sold at a fraction of the predicted market price (Cr$0.30 instead of Cr$2.50 per kilo) because it coincided with that of the major onion-producing area of the North-East in Pernambuco, the neighbouring state. This resulted in a considerable loss of income for the colonists. Poor planning was entirely responsible for this minor disaster because the feasibility study had earlier recommended quite specifically that the crop should be marketed in the inter-harvest period.[26] A similar experience was had with an experimental tomato crop in the same year.

Marketing crises have also been caused by the over-dependence of projects on a particular crop, whether traditional or luxury. In Sumé the processing factories (Peixe and Rosa) which bought the project's tomato production, suddenly cancelled their contracts in 1973. For a while there were fears of an economic disaster but a crisis was averted when a contract was eventually signed with another plant (Cica Norte), which now takes 90% of tomato production. In São Gonçalo a similarly disastrous situation seemed imminent when the factory in Arcoverde (Pernambuco) which purchased 90% of the banana crop refused to buy a particularly large harvest. Only after considerable searching and the direct intervention of the central DNOCS administration in Fortaleza was a buyer found.

One of the main causes of the haphazard nature of marketing procedures on irrigation projects is that DNOCS has made no systematic attempt to coordinate the activities of irrigation projects in the field of commercialization. GEIDA proposed, as we have seen, a series of local centres responsible for collecting, grading and marketing farmers' crops, thereby diminishing costs and increasing profits for the actual producers.[27] In reality, however, there is no such coordination amongst DNOCS schemes, either locally or regionally. Each project appears to evolve its own production and marketing pattern based on a compromise between on the one hand which cultivation system is considered technically and organizationally feasible by the project administration, and on the other what outlets for such crops the hierarchy can secure. There is no concerted attempt to coordinate the production patterns of projects so that particularly lucrative markets may be efficiently exploited. One such attempt was in fact made to organize a form of collaboration between the producers' cooperatives of projects in the state of Paraíba, notably Sumé and São Gonçalo. However, their totally different agricultural plans, combined with practical problems of communication, resulted in the collapse of this short-lived experiment.

Another handicap at the marketing stage which decreases potential profits lies in the fact that irrigation projects do not yet have the facilities for processing and industrializing the crops they produce. Morada Nova has managed to partially circumvent this problem with cotton, its main money-earner, by sending it to a central cooperative (privately-owned) in Fortaleza (*Cooperativa Central de Beneficiamento*). The project cooperative recently made a special arrangement whereby 30% of the profit obtained from processing the cotton will be returned to the scheme. No such arrangements are possible for rice, however, most of the profits from which accrue to the intermediary wholesalers and the processing factories. Bananas and tomatoes are also sold at relatively low prices, most of the profits going to the canning factories. Until DNOCS manages to set up its own processing plants or make special arrangements with private factories, the income-earning potential of these more traditional crops will always be limited. This will probably only be economically viable once production has reached much higher levels than at present.

One of the ways in which DNOCS hopes to aid the efficient functioning of its irrigation projects appears to be by means of promoting what is calls 'cooperativism'. Although the agency has not made any official commitment to this effect, it lays considerable emphasis on encouraging *cooperativismo* in its publicity material. Nevertheless, it does not define precisely what it means by the term. Generally speaking, however, it refers to a vaguely defined spirit of collective cooperation among colonists which is, ideally, conducive to greater harmony and economic success. Simultaneously, *cooperativismo* also implies that irrigation farmers should follow unquestioningly the guidelines set down by the administration, secure in the knowledge that such policies are in their best interests.

In theory at least, DNOCS tries to select candidates who are thought to have a predisposition towards 'cooperativism'. During the final stage of the selection process interviewers try to assess qualitative personality traits of applicants such as their 'degree of initiative', 'receptiveness to change', 'sociability' and their level of 'cooperativism'.[28] These qualities are graded, along with other qualifications such as age, experience and family size, and a grand total of points is arrived at which is used as a basis for choosing colonists. In practice the system is clearly full of flaws, especially the very simplistic attempts to classify personality characteristics and grade them numerically. Even in its most sophisticated form such a process is open to serious criticism, and DNOCS staff, however good their intentions, have neither the ability nor the time to undertake such a task efficiently. In fact many selectors realize the inherent futility of such attempts and classify aspiring colonists more according to their other more easily identifiable qualifications, although this is frequently supplemented by superficial impressions about personality. It has been noted that, as a result, more extrovert candidates have sometimes received preferential treatment on the basis of their apparent ability to communicate more easily.[29]

DNOCS is liable to regard potential deviation from the 'cooperativist' ethos in an unfavourable light. During the selection interviews DNOCS staff try to ascertain whether candidates are willing to accept the administration's authority and guidance. In fairness, it should be said that interviewers also look for some justification from an applicant as to why he would follow advice given him, rather than taking his blind acceptance of authority as good enough. The point is, however, that DNOCS does not appear to assume that farmers may have constructive contributions to make to the project which might differ from accepted, official policy. This intolerance of suggestions from below, detectable even at the selection stage, strongly contradicts any pretensions towards a 'cooperativist' spirit, however loosely defined.

A significant illustration of how the agency reacts to threats, real or imagined, to its monopoly over decision-making on projects, was provided in Morada Nova. The application of one farmer had been ignored by DNOCS for three years because of allegations that his political record made him a potential danger to the 'smooth functioning' of the project. The person in question had been secretary of the Federation of Rural Workers of Ceará, subsequently 'intervened' by the military government. He was later elected leader of the Rural Workers' Syndicate in the nearby town of Limoeiro do Norte. Eventually he gained admission to the project by pre-empting attempts to delay consideration of his application so that he could be made ineligible on the grounds of his age. This he managed with strongly worded letters of his own, supported by statements that he had committed no criminal offences as a union leader. Reluctantly, the administration accepted him, and it now seems that he is a 'model' irrigation farmer.

Once colonists have been selected and installed on their plots, there are two main areas where DNOCS, in theory at least, tries to encourage a 'cooperativist' spirit. In the work area this is done through the cooperative, which is responsible for distributing credit to project members and for commercializing their produce. In the non-work field attempts are made to promote community development with the participation of families in group meetings. In fact, the department responsible for the social welfare of colonists is called the 'Service of Social and Cooperativist Promotion' (*Serviço de Promoção Social e Cooperativista*).

Apart from its commercialization and credit functions a project cooperative is supposed to stimulate 'cooperativist education' and encourage colonists to 'place collective interests above individual interests'.[30] The cooperative is run by an administrative council which consists of five elected members: a president, vice-president and secretary as well as two advisers from the project administration. Members meet once a year at the AGM to discuss general policy, carrying out elections of new representatives every two years. The AGM is virtually the only opportunity which colonists have to exert any influence on policy formation. They do meet briefly every two weeks or so to receive their credit allowances and to hear any special announcements by the *gerência*. At these meetings, which would be a good opportunity to sound out popular opinion on various matters of concern to all colonists, the emphasis is placed very heavily in the opposite direction, the issuing of instructions. Criticism of existing policies is neither invited nor encouraged in São Gonçalo or Sumé. In Morada Nova, credit is distributed in small batches to groups of colonists, who do not have these regular meetings.

One of the main handicaps towards developing such forms of communal expression stems from the sheer size of projects as they grow and the geographical dispersal of families. In São Gonçalo, for example, 77 of the 123 colonists had been housed in two nuclei, while the remainder lived in isolated huts scattered both within the irrigated zone and outside of the hydraulic basin altogether. In Sumé this was no problem given the smallness of the project. As schemes expand to their full size, however, limits are bound to be imposed upon the regularity and significance of such meetings.

The whole system of irrigated agriculture in the form of individualized, family plots, combined with very rigid planning from the top down, conflicts strongly with DNOCS's official preoccupation with encouraging *cooperativismo*. The ideas behind these two concepts are basically self-contradictory. From the moment a colonist enters a project he is told exactly how much land he should cultivate, what crops he should grow and in what quantities, what treatment to apply, when to harvest and to whom he should sell his crops. A high degree of direction is to be expected given the novelty of irrigated agriculture for recruits and the marketing constraints, but at the same time almost no room is left for individual initiative or free choice. The farmer is not expected to make his own decisions regarding production, nor is he encouraged to make suggestions as to possible alternative methods. After all, he has been admitted to the project on the assumption that the administration and agricultural technicians know best. In practice, of course, the farmer often does have to use his individual initiative, albeit unofficially. I would suggest that colonists frequently have to make do with their little acquired knowledge when assistance from *práticos* is not forthcoming. Another practice which illustrates the need for farmers to exercise individual initiative in the private marketing of crops out of economic necessity, already mentioned in chapter 5.

A typical illustration of the absence of collective discussion behind decisions taken by DNOCS administration was provided by a meeting of the Sumé project director with the colonists. DNOCS had decided that it would be a good idea to introduce a compulsory savings scheme to set aside 40% of colonists' net profits at the end of the year. Half of this would be pooled to pay for additional labour, obviating the need for bank loans, which charged interest (although at a negative rate, in fact). The remainder would be deposited with the Bank of the North-East (BNB). While his audience listened in silence, the director tried to get across the advantages of saving, both for their families' sakes and as a contribution towards 'national development'. He repeated several times the idea that *nordestinos* were not used to saving but that this must change if they were to become successful colonists. It was no use hoarding cash or buying cars. One or two colonists, who had bought jeeps at considerable subsequent expense, were cited by the director as examples of bad spenders, their search for 'status' having overcome prudence. How much better it would be, he argued, if the surplus were invested into buying vehicles which could be used to transport their crops for marketing instead of relying on DNOCS. He took great pains to stress the need for the farmers to break with the old ideas of paternalism, of doing away with their dependence on the generosity and help of the administration. Some modifications were made to the details of the savings scheme after a brief discussion involving the director and four colonists of the twenty-four present. At the end of the meeting the director rebuked the majority for their lack of participation, stressing the need for collective decision-making. The savings plan would start in the near future. No vote was taken.

The second area of 'cooperativist' action comes under community development. This is the responsibility of social workers, whose job it is to hold regular group meetings with housewives, to encourage discussion of common problems and to communicate such feedback as there is to those in charge of planning and administering the project. Within each housing community the social workers try to hold regular, monthly meetings to discuss various aspects of life on the irrigation scheme such as health and sanitation problems, vaccination, purchase of water-filters, their children's education and cost of living problems. The social arm of DNOCS does appear to be making a genuine attempt to promote group discussions and the open expression of opinion.

An illuminating insight into the potential usefulness of such meetings was provided by my attendance at a discussion amongst a group of housewives in São Gonçalo, concerning a recent decision taken by the directive council of the cooperative. This prohibited outright the keeping of goats, dogs, cats and poultry on the housing nuclei. As with the failure to provide families with subsistence plots, bad planning had resulted in houses not having small, fenced enclosures where such animals might be kept. Families need goats' milk and chickens' eggs to supplement their poor diets, and they reacted by keeping animals running loose in the housing settlements, an annoyance to some and a health hazard to all. The decision to forbid the keeping of animals, taken without consultation of project families, was greeted with considerable hostility by the meeting and, as a result, the matter was going to be reconsidered by the cooperative. The social worker suggested that the problem could be solved by each family building its own enclosure. This, in turn, brought protests that their husbands had no time to spare given the heavy work load involved with keeping an irrigated plot, nor the money to buy wood and wire fencing materials. The meeting closed with the suggestion that they might cooperate with each other to build the enclosures.

Social workers, then, constitute the only real channel through which the interests and everyday problems of colonists and their families may be communicated to the DNOCS administration. Unfortunately, the concern shown by social workers for promoting genuine cooperation amongst project members at the grass roots is not mirrored at higher levels of the administration, where lip service seems to be the only contribution. At a recent regional meeting of the social work division of DNOCS irrigation projects, the coordinator from central administration deemed the matter of 'cooperativism' unworthy of comment because of a 'lack of concrete data'.[31] Project administrations are designed to issue orders to colonists rather than listen to their points of view. Instead of exerting collective pressures in support of their own interests, irrigation farmers have to be content with expressing their opinions in a highly fragmented, sporadic and indirect fashion.

The aim of increasing agricultural production, thereby generating higher incomes for irrigation farmers, has, therefore, met with very limited success in relation to the optimistic forecasts made by feasibility studies. At the same time, 'cooperativism' as a means of aiding this process exists more in the pages of DNOCS brochures than in the everyday activities of irrigation schemes. This is, I suggest, principally because the hierarchical way in which the agency is organized is fundamentally antagonistic towards the development of a genuine cooperativist spirit amongst farmers and administration.

7. Alternatives to irrigation

From the preceding analysis of government irrigation schemes it is patently obvious that they have made very little contribution towards solving the deep-seated agrarian problems of North-East Brazil. More importantly, though, existing irrigation strategies have little potential to alleviate these problems, no matter how large the scale of this exercise in rural development, which is so costly both in financial and human terms. Economic returns are low in relation to the vast capital investments, as well as being unequally distributed among the handful of colonization farmers.

However, it is in other aspects of the policy where the harmful effects on the poor rural population of such 'development by force' are more tragically apparent. The forced migration of peasant families from their valley terrain, which has been so prominent a feature of DNOCS tactics during the past seven years, is now being adopted as a permanent feature of irrigation policy over huge areas in the region's largest river valley, the São Francisco. Here, the government agency responsible for irrigation, CODEVASF, has been obliging thousands of rural families who have for generations been enjoying a stable existence, to leave the area and find work elsewhere. No policy could be more guaranteed to encourage rural–urban migration.

The violence which accompanied the dispossession of peasants in DNOCS project areas is currently being mirrored in the São Francisco valley. In the Propriá region in the state of Sergipe, peasants dissatisfied with the amount of compensation offered and reluctant to leave their homes have been intimidated by the local police and threatened with violence. Such tales of illegal pressures brought to bear on a defenceless population are common in areas where irrigation schemes are being implemented. On the São Desiderio project run by CODEVASF in the state of Bahia, farmers were allegedly tricked into selling their lands with false promises of irrigated plots on the new scheme.[1] In 1976 these irregularities reached such a peak that the local bishop made a formal protest to the Minister of the Interior. In the face of increasing popular discontent in areas affected by irrigation schemes, the actions of CODEVASF in applying illegal pressures were condemned by the Pastoral Land Commission of the National Conference of Brazilian Bishops (CNBB) at a regional meeting held in the São Francisco area in October 1976. Little room is left for the sort of optimism which led one major Brazilian magazine to herald irrigation as 'a solution for unemployment' in the North-East.[2]

If the government is serious in wishing to implement policies of rural development which will help to stabilize the agrarian population of the North-Eastern interior during drought periods, alternative longer-term strategies are necessary. These policies, whether they replace or complement irrigation in its present form, would have to ensure that the specific rural groups most at risk to the rigours of the drought are helped in ways which increase their resistance. Current strategies manifestly do not do this. Planners should take into account the broad, socio-economic nature of the drought phenomenon, identify those groups most in need of direct assistance on this basis, and provide the vulnerable population with the means to acquire economic resources that enable poor farmers to build up food and cash reserves to tide them over difficult periods such as those of acute rainfall shortages. Complementary improvements in domestic water supply would also help to discourage migration. This goal can be achieved either with or without irrigation but the fundamental point is, to repeat, that any anti-drought strategy must reach those at risk and not benefit the already privileged.

DNOCS's irrigation programme is not the only example of a rural development strategy which tends to improve the lot of those farmers who already enjoy relatively better positions. In the São Francisco valley it also appears that irrigation schemes have a similar selection procedure designed to pick out those farmers considered to have the skills necessary for irrigated agriculture. Successful applicants are usually those in better health, with more capital and who demonstrate more of the individual, entrepreneurial initiative needed for the family plot system. With a rejection rate of 90% on the Bebedouro project, for example, CODEVASF has plenty of scope for careful selection of its recruits.[3] Medical and aptitudinal tests eliminate many small farmers who are precisely those most likely to be affected by the crisis conditions such as drought. In North-East Brazil as in other parts of the world, 'supplying irrigation water is very much a matter of "to him that hath shall be given" – it is most valuable in the hands of the already wealthy'.[4] Unfortunately, the cost-benefit approach to irrigation adopted by GEIDA, DNOCS and CODEVASF will probably always mean that the narrow criterion of economic success will prevail above considerations of social need.

Chapter 5 also showed that the income derived from irrigated agriculture in the North-East tends to become concentrated in the hands of a small minority of colonists. Once again, this is certainly not a new phenomenon and has been observed, for example, in the irrigated lands of Africa and Mexico as well as in Asian 'green revolution' rural development. Even on what is considered to be the world's largest and most successful irrigation project, in the Gezira region of the Sudan, the incomes and life-styles of tenant farmers have become sharply differentiated. The more prosperous, with consumption patterns to match their new status,[5] regard themselves as 'gentlemen managers' rather than rural workers. In contrast to the Brazilian experience, however, this scheme has benefitted millions of rural producers and covers 720,000 hectares irrigated between 1958 and 1962. It is the cheapest irrigation project in the world at £20 per acre and tenants' incomes the largest for comparable environments.[6] Millions of nomads and semi-nomads are also being benefitted by supplementary development schemes.[7]

Another example is provided by the Tepalcatepec river basin development of Central Mexico where the government has irrigated and provided the infrastructure for 64,000 hectares of land, and which yields a 13% return on capital investment. Differential access to credit has meant, however, that a handful of large, private farmers monopolized production and income, while the mass of small landowners and *ejidatarios* progressed very little.[8] Feder has also observed that in irrigated areas of Mexico improved agricultural techniques based on sophisticated technology have resulted in the benefits of increased production accruing to small groups of rural landowners and non-rural industrialists. While the export of luxury crops produced under irrigation has been encouraged, investment into peasant agriculture and land reform has declined. Under the new 'green revolution' technology, it is argued, labour is replaced by machinery, and terms of employment on modern farms, which require seasonal manpower, become worse. These factors have been responsible for increasing mass poverty, rural unemployment and social unrest in Mexico.[9]

Other studies of the effects of technological improvements on traditional agriculture suggest the emergence of similar trends, the beginnings of which are already clearly observable in DNOCS's irrigation programme. In Colombia, for example, the introduction of new methods of coffee cultivation benefitted the already privileged. In this case, as in the Mexican example, farmers with the most land and capital were able to secure credit in order to purchase higher-cost inputs of new varieties and fertilizers necessary to obtain higher incomes.[10] In India a small minority of farmers prospers through cultivating high-yielding crops but the new agriculture, which is capital-intensive, does not help to relieve the problem of massive seasonal unemployment, thus exacerbating mass poverty in rural areas in that country.[11] In all of these examples it appears that social needs of the wider agricultural populations affected by such technological changes have been largely ignored. The pattern of production has been determined by factors such as profitability, export markets, foreign exchange earnings and productivity rather than consideration of the nutritional needs of the population or the distribution of higher incomes. Assumptions that somehow the benefits of improved technology will 'filter down' to the poorer sectors have not so far been realized.

Any alternative anti-drought strategy in North-East Brazil should respond more directly to the central causes of rural unemployment, poverty and migration. As a starting point, therefore, one possibility lies in the modification of existing irrigation strategy in order to correct some of its basic faults. After all, considerable amounts have been invested in irrigation and it seems reasonable to suggest changes which might enable projects to assist groups particularly vulnerable to the drought, thus capitalizing on government expenditure made so far.

The individual, family plot system might, for example, be replaced by a more collective form of cultivation. It will be remembered that one firm of consultants, Hidroservice, emphasized several times in its reports to DNOCS that the only way of eliminating the inequalities and unfairness inherent in dividing the land into separate plots was to cultivate the whole area as a single unit using salaried labour. The justification for their advice is reflected in the extremely skewed income distribution amongst colonists, caused to a large extent by differences in soil quality. Several schemes in Africa, for example, have mechanized farming with paid labour and no settlers. The Richard–Toll scheme in Senegal producing rice and the Tendaho project in Ethiopia for cotton are both situated in dry areas and show that such a system can function efficiently.[12]

The whole system of irrigated production on DNOCS schemes emphasizes individualism: the division of the land, of individual 'entrepreneurial' responsibilities and of income. High-income colonists are held up as examples for their less successful colleagues to follow. Yet such a system is divisive amongst farmers and is itself antagonistic towards any attempts at encouraging cooperation. Studies of irrigation in other parts of the world have shown that sometimes the small plot system leads to conflicts amongst farmers and greater economic polarization.[13] Conversely, equalization of access to factors of production under irrigated agriculture has been found to diminish such conflicts.[14]

Collective cultivation would in theory have several advantages over existing methods. The land would not have to be sub-divided, each plot with its own particular cultivation pattern requiring individual guidance and supervision. The land could be prepared and cultivated in larger units, diminishing production costs and losses through individual inadaptability. Wage-payments could be related to productivity and all produce marketed through a cooperative. Most importantly, from an anti-drought point of view, larger areas could probably be brought under production and more labourers per hectare absorbed, provided that production techniques did not become too capital-intensive. Rural workers would not have to undergo an expensive and time-consuming selection procedure, allowing poorer sharecroppers and smallholders from drier areas of *caatinga* to be absorbed more easily.

It would, in theory, be possible to irrigate humid valleys in the North-East without the expropriations and displacements typical of current DNOCS strategy. In the Mysore region of India 309,000 hectares of river valley lands have been irrigated, resulting in the transformation of a subsistence agriculture into a cash economy, thereby increasing incomes and living standards, as well as integrating the region into the national market.[15] No expropriations or displacements took place; irrigation facilities were provided by the government and superimposed upon the existing property structure, which consists exclusively of smallholdings below 37 hectares. Irrigation did tend to reinforce economic differentiation but overall incomes were raised and there are certain inbuilt obstacles to the accumulation of individual wealth (such as inheritance laws, the disappearance of the joint family and increased population size) which have resulted in the benefits of new technology being perhaps more evenly spread than in many irrigated areas of the world. In addition, 'dry' villages outside the irrigated zone have developed economically by providing secondary and tertiary services to 'wet' villages, thus sharing in the regional prosperity.

The situations in Mysore state and North-East Brazil are difficult to compare directly since the pattern of development subsequently established in the former region was influenced by the caste system, the structure of economic values and receptivity to economic change. As far as the present study is concerned, however, the one important lesson which may be drawn from the Indian example is that it is possible to successfully irrigate land without the need for modifying the property structure or for elaborate and time-consuming selection procedures which tend to eliminate poorer, less qualified farmers who are those most in need of assistance. If such a system were tried in Brazil, however, it would have to be in areas where the distribution of landholdings was fairly even, minimizing the possibility of wealth accumulating in the hands of a small minority. Where DNOCS originally tried this system, such as on the São Gonçalo project from the 1930s onwards (discussed in chapter 3), income was highly skewed because of the heavy concentration of landownership.

However, in order to make a significant impact on the population most vulnerable to the drought, irrigation projects would have to be installed not in the relatively fertile, densely populated valleys as at present, but on the dry *caatinga*. This would avoid many of the social pitfalls currently associated with expropriation and displacement of valley populations, while at the same time ensuring that those poorer farmers with fewer resources would be the first to benefit and not the last, as is currently the case. The initial costs of pumping water to higher areas might be greater, but there would be fewer delays caused by muddy and flooded valleys during winter months. A collective system of production would, furthermore, obviate the need for selecting farmers with special skills or abilities, providing stable employment for all those with a commitment to the project.

A move away from family plots to a collective system of land tenure and agricultural production is, however, likely to be resisted by DNOCS for several reasons. Firstly, agency staff argue that the *nordestino* is an individualist, unsuited to any type of collective system. One factor which suggests that there is more room for such cooperation amongst small producers than DNOCS cares to admit, lies in the tradition of *mutirão* or mutual self-help amongst peasants.[16] A second reason which favours individualized units, one suspects, is fear of political mobilization. Although DNOCS pays lip service to the principle of cooperation in the field of working activities, in practice it does little to encourage a true collectivist spirit. It may well be that DNOCS views the collective organization of labour as a challenge to its authority.

It has been argued that the sort of paternalistic attitude which DNOCS adopts towards its colonists, denying them any real participation in decision-making, can have a positive effect. Galjart suggests that local development agencies should assume the role of patron, thus preventing a relapse into the traditional, boss-dominated social structure and mediating with higher levels in the patronage system.[17] Although it has to be admitted that tenant farmers on irrigation projects appear to be as dependent upon the government as they were on their old, private landlords, such a tendency should not be accepted as inevitable. Official paternalism in the form of strong direction from the top down perpetuates what Erasmus calls the '*encojido* syndrome', the continued dependence of the peasantry on officials rather than the autonomous decision-making characteristic of true voluntary organizations.[18] In the Gezira, for example, an independent tenants' union sprang up as a reaction against official paternalism, and demanded increased participation in the control and management of the project. Fundamental issues such as cropping methods, sales and village improvements are discussed.[19]

Even without making drastic changes, DNOCS could take steps which would ameliorate some of the obviously negative effects of current irrigation strategy. Small farmers who qualify for little compensation when their property is expropriated, or none at all if they cannot prove *de jure* ownership, should receive fairer treatment. Ideally, the expropriation laws should be amended to give planners the authority to see that the dispossessed have sufficient resources to re-establish themselves elsewhere. In the case of non-landowners in particular, a relocation policy is badly needed which would rehouse and re-employ the thousands of families forced off the land. In this respect CHESF has set a precedent in the São Francisco valley by giving those families dispossessed by the building of the giant Sobradinho dam the option of moving to new towns constructed nearby. Apart from improving monetary compensation and relocation, another potential improvement lies in associating outsiders with irrigation schemes. In the Gezira, for example, nomad farmers are allowed to pasture their livestock on the project.[20] On DNOCS schemes perhaps non-member farmers could be granted access to water supply facilities.

Another major area for improvement is in the commercialization of produce. One of the principal causes of projects' limited income-earning capacity in the North-East is their inability to produce and market high-value crops, despite attempts to do so. Every large project should have a pilot scheme to evaluate the suitability of proposed strategies. The success of the Gezira scheme has been partially attributed to careful advance planning.[21] In the Kafu project, Northern Rhodesia, 173 hectares were carefully studied and experimental farms established to consider alternative cropping and irrigation methods before the major scheme was started.[22] In the Medjerda project, Tunisia, advance production and marketing studies enabled artichokes, an extremely high-value crop, to be successfully produced on a large scale.[23] DNOCS's own Sumé scheme has achieved similar results with tomatoes, albeit on a far smaller scale.

There is, therefore, the scope for expanding production of fruit and vegetables on irrigation projects in the North-East, even if predictions of market capacity have been over-optimistic. The onus rests on DNOCS to exploit these possibilities. In the São Francisco valley private companies have shown that grapes and other luxury crops can be successfully grown and marketed.[24] Irrigation projects might also benefit from recent government proposals concerning the establishment of new centres for processing high-value produce such as tomatoes, cotton, pineapples and other fruits.[25]

Anti-drought strategy need not necessarily, however, be based on irrigation. If the drought problem consists essentially, as defined here, of impoverished rural groups whose members living at the economic margin have few resources to see them through crisis periods, there are a number of other possibilities which the government could explore. These policies would be aimed at improving the lot of vulnerable sectors, such as sharecroppers and smallholders, so that they would not be forced to migrate except when the most serious of droughts strike. These strategies would increase cash incomes and food stocks by such means as better farming technology, agricultural credit, cooperatives and possibly land redistribution. Although not as 'glamorous' as irrigation, these alternatives have several potential advantages. They would be far less costly to implement, requiring none of the expensive and time-consuming engineering work needed to build irrigation schemes. A far greater number of people could be easily and more quickly reached since there would be no laborious selection procedure. Nor would those in most need be excluded on medical or aptitudinal grounds.

Much more efficient use could be made of river valley lands without recourse to expropriating and irrigating large sectors. This is especially so in perennial river valleys such as the São Francisco, as well as in those where dams regulate the water flow so that farmers receive a more balanced supply all year round. Even with no additional inputs, banana yields in some non-irrigated valleys are comparable with those on DNOCS projects. In the Moxotó–Gorotuba and Piranhas–Açú valleys yields of 45 and 47 tons per hectare are common[26] compared with the maximum figure cited by DNOCS of 55 tons (see table 17). Studies have suggested that the productivity of many food crops could be doubled or tripled using more modern methods.[27]

A recent analysis was made of the probable effects of introducing more modern techniques in two areas of the *sertão*, Serra Talhada and Brejo Santo. The study concluded that if manual labour was supplemented by animal traction, use of fertilizers and disease control, agricultural production could be increased by between 50% and 100%. As the result of larger areas brought into production, more labour would be employed.[28] Larger property owners would tend to benefit more because of the greater quantity of land at their disposal, but on balance, it is suggested, incomes of all groups would be increased in absolute terms.

In order that smaller farmers should be able to take advantage of more modern inputs, a complementary programme of agricultural credit would be needed. At present rural credit is monopolized by large landowners, who are able to offer collateral to the banks. Even emergency drought credits, as chapter 2 showed, are monopolized by the bigger establishments whose resources in the form of cattle offer security, both for the landowner and the lending institutions. As soon as a drought seems imminent, banks call in loans to smaller farmers, increasing further still their vulnerability. The resources of many sharecroppers and smallholders are whittled away by the obligation to receive credit privately at punitive interest rates, as well as to sell their crops at prices well below the market value.

A system of small-scale loans is badly needed, therefore, which would break this vicious circle and the monopoly over credit and commercialization arrangements held by private landlords and merchants. What is required is not a general increase in levels of credit for all farmers, but a system of allocating loans at preferential interest rates to specific groups most in need. If there were no discrimination along these lines, the danger would exist of agricultural loans being used for other purposes such as land speculation. The demand for agricultural credit could be artificially stimulated, reducing the amount available for small farmers and increasing the importance of political influence as a means of securing loans.[29]

The apparently 'primitive' methods of agricultural production used by sharecroppers in the *sertão* represent rational responses to the limitations imposed upon their behaviour by their economic and political environment.[30] It follows that any new system of agricultural technology or credit can only be implemented successfully if such methods are demonstrably profitable for the recipients. Government authorities cannot legitimately present blind adherence to 'tradition' on the part of peasants as an excuse for the failure of rural producers to follow official policies.[31] It is up to the planners to devise strategies which make innovation profitable and risk-free for farmers. Increased risks which accompany new credit schemes are, for example, a common cause of default on loan repayment because inadequate attention has been paid to marketing, prices and other support facilities.[32]

Such support may be provided by cooperatives and similar bodies which help small producers to organize, increase their bargaining power and bypass the long chain of intermediaries which absorb a large proportion of the crops' marketable value. In the North-East there are approximately 650 cooperatives, but few of these make any impact upon the monopoly over commercialization held by the network of traditional intermediaries.[33] There are isolated examples of producers' cooperatives which manage to become independent of landlords and merchants. In the *sertão* of Aratuba, Ceará, for example, a large farm was bought by the parish and the land redistributed to sharecroppers of the region who work it commercially and are purchasing it collectively on a mortgage basis. The communal cotton crop is deposited with a local cooperative to await inter-seasonal price increases, when it is sold.

Generally speaking, however, small producers in the *sertão* are not organized and the cooperatives which do exist suffer from serious defects. Their administrative personnel are poorly trained and local political patronage prevents their reaching a broader clientele. Their small scale of operations and lack of financial resources or public backing are further obstacles. Furthermore, they deny any participation in policy formation or decision-making to the agriculturalists themselves.[34]

One possible way of helping small and medium-sized producers to receive a larger share of their crops' value lies in the setting up of cooperatives on a state-wide basis, or *Companhias Integradas de Desenvolvimento Agrícola* (CIDAs). These centres resemble in many ways those envisaged in the GEIDA proposals for irrigation farmers but never implemented. They would, according to SUDENE, transport farmers' produce and provide grading and storage facilities as well as process crops. Technical assistance would be given to farmers, credit made available and the produce marketed directly by the CIDAs. Run by professional administrators, these centres would be situated in state interiors and have large-scale operations, involving participation of producers in the fixed capital of the enterprise.[35] One of the practical limitations, and one of which SUDENE is fully aware, is that CIDAs would have to be located in areas where small producers are free, or could free themselves, of obligations to landlords, enabling them to choose through which channels they wish to market their crops.[36]

An additional means of assisting small producers to maximize the economic return on their crops, whether marketed through cooperatives or otherwise, is a system of minimum prices. Since the early 1960s in North-East Brazil, official attempts to create a structure of minimum prices for farmers have been made through the *Grupo Executivo de Modernização do Sistema de Abastecimento* and more recently with the *Sistema Nacional de Centrais de Abastecimento.*[37] It was intended that commercialization centres should be established in the largest cities to receive produce, organize buying and selling, provide farmers with technical services and guarantee minimum prices to them. Such institutions exist in Recife and Fortaleza but their overall effect is marginal, especially in relation to the distant *sertão*.

Apart from using these centres farmers may borrow from the Bank of Brazil, which acts as agent for the *Commissão de Financiamento de Produção* (CFP). There are two credit systems in operation. Under the first, farmers may obtain 70–80% of the value of their future harvest, calculated on the basis of guaranteed prices. In this case the producer may sell part or all of the crop within the loan period as long as he covers his debts to the bank. Once the 180-day loan period has expired, the crops do not pass automatically to the CFP. Under the second type of arrangement, farmers may receive 100% financing but if debts are not paid within the time limit set (90–300 days), the crops, which are held in government bond, pass into CFP hands.[38]

These programmes, however, do not at present benefit significant sectors of the rural population but have remained extremely small-scale.[39] It is possible, nevertheless, that large numbers of small producers in the *sertão* might take advantage of minimum price supports if such a system could be made to operate through an institutional framework such as the proposed CIDAs under the authority of SUDENE. This would be one way of helping to break traditional landlords' monopoly over credit and marketing, and provide many potential drought victims with extra financial support. In order to be effective, however, considerable efforts would have to be devoted towards reaching those in need of assistance, rather than letting the already better-off rural groups monopolize new facilities. This implies a process of selection which may well prove difficult to undertake because of political constraints. It seems unlikely that powerful groups will allow significant privileges to be granted to poorer producers while they themselves are denied access to such concessions, unless reforms are backed up by authority from the highest levels.

Land redistribution, by providing credit for land purchases by small farmers, is another method which could strengthen the resistance of potential *flagelados.* It has been shown that in the North-East small establishments cannot fully utilize their family labour because of the limited land and other resources available. Indeed, the ability of small farms to absorb this 'excess' labour, as chapter 2 showed, gives them a certain strength and ability to support dependents in the face of scarce employment opportunities. One study of two *sertão* municipalities has suggested that if small farmers acquired additional lands the value of their production and, consequently, net incomes would be increased substantially.[40] In order that new lands could be effectively utilized, however, credit would also have to be made available for the purchase of other inputs such as seeds and fertilizers.[41]

The biggest difficulty would be actually obtaining the land needed by the farmers. This might necessitate expropriations from larger landowners, which may or may not be politically feasible. DNOCS has set a precedent with its activities in this field, however, so that in theory expropriation is possible. Secondly, the quality and quantity of land would have to be balanced to ensure that land and labour could be most efficiently used.

Finally, the size and composition of the family labour force would have some bearing on the size of establishment created in order to ensure full employment.[42]

In spite of the potential advantages, however, the government's record on land purchases for smaller farmers is not encouraging for the future. The PROTERRA programme, for example, started in 1971, was designed to provide subsidized credit for the purchase of modern inputs and land to be redistributed to small and medium-sized landowners. However, nearly all of this credit has been used for large cattle and sugar estates, a portion having been channelled into short-term, speculative and non-agricultural purposes. Landless workers have been totally omitted from the scheme because they cannot offer guarantees to the banks. Small landowners are in a similar position since their properties are of little marketable value and they have nothing else to offer as collateral.[43] By 1974 in the North-East only seventy-four land titles had been distributed to small farmers.[44]

The possibility of establishing more family farms in the future is also likely to be undermined by the official policy of providing credit for the setting up of large cattle estates. Under SUDENE's 34/18 fiscal incentive scheme, 90% of loans to agriculture are used for livestock enterprises.[45] These projects, far from absorbing excess labour, are capital-intensive and tend to create more unemployment by displacing semi-subsistence farmers from the lands taken over with the aid of official credit. In Maranhão, Bahia and Piauí cattle estates averaging 35,000 hectares have been established, reminiscent of the trend towards huge *latifundia* in the Amazon region. It has been estimated that if 34/18 funds were applied to land redistribution, 175,000 families could be absorbed, creating 350,000 jobs.[46]

Colonization of previously unsettled areas by rural families from the North-Eastern interior is another strategy which has been given much consideration by Brazilian planners. The first official mention of colonization specifically in relation to the *sertão* and its drought problem was made by the GTDN report which saw in it 'a means of absorbing the excess population of the semi-arid zone and producing foodstuffs which would help feed that region'.[47] In the early 1970s renewed emphasis was made on colonizing the Amazon region, especially in new rural communities (*agrovilas*) along the Trans-Amazon Highway. This was formalised in the POLAMAZONIA plan under the Second National Development Plan in 1974.

The main justification for colonization has rested on the fact that the North-East is considered to have an excessively large rural population which the region's limited resources cannot support. A recent study calculated that by 1980 the *sertão* would have an excess, low-income rural population of 952,000.[48] It is possible to debate whether the concept of 'excess' population is justified in the context of a largely semi-subsistence agriculture in which the family sector absorbs a large proportion of the labour force. Nevertheless, there can be little doubt that the strain imposed on resources by the large population, especially during drought crises, exacerbates the problem of poverty, unemployment and rural–urban migration. Although by no means a total solution to the drought problem, selective colonization appears to offer a partial answer, particularly in crucial areas where the pressure on land makes the local population especially vulnerable to the drought. Ideally, a greater redistribution of under-utilized land within the North-East would serve the same ends, but this is unlikely to be politically feasible on any significant scale. Otherwise, according to one estimate, perhaps over half the 'excess' population could be resettled within the North-East itself, in Maranhão, Bahia and Piauí, the remainder going to Amazônia.[49]

One advantage which colonization offers over, say, irrigation is that it is less costly. Several studies of non-irrigated colonization suggest that the investment costs incurred in settling one family are one-sixth those on irrigation schemes.[50] Non-irrigated colonization, of course, absorbs labour, whereas irrigation projects tend, on balance, to displace rural workers.

In spite of the theoretical advantages of colonization, however, several reservations should be borne in mind. Firstly, the availability of uninhabited areas for such projects may well be limited by the PROTERRA and 34/18 schemes, already discussed, which tend to encourage the establishment and perpetuation of *latifúndia* which are basically capital-intensive and devoted to livestock production. It is precisely in the more fertile areas of Maranhão, Bahia and Piauí, currently viewed as regions for settlement by families from the arid interior, where cattle estates are springing up. Secondly, many migrants from the *sertão* have suffered harsh treatment at the hands of unscrupulous employers, truckers and boarding houses, a system which has been labelled as 'motorized slavery' by its more extreme critics.[51] If such abuses are widespread they call into question the whole value of colonization. However, if relatively infrequent, as some observers maintain,[52] they do not detract from the general value of colonization as a means of reallocating human resources. In the ten years that colonization has received official encouragement, however, only 200,000 hectares have been settled and 5,000 families absorbed.[53] Unless it receives more government support along the lines envisaged by d'Apote, its impact on relieving economic pressures in the *sertão* will be very limited.

The improvement of agricultural productivity in drier areas of the *sertão*, as opposed to the more humid valleys, is one strategy whose potential has so far been largely ignored. Two areas where such improvements might increase the drought-resisting capacity of the rural population are cotton and cattle production, activities which are already resistant to rainfall shortages. This resistance could, however, be greatly increased if more efficient use were made of the land and other resources available. The productivity of aboreal cotton (*mocó*), the variety traditionally cultivated in the interior, could be greatly increased by the careful selection of seed varieties and the application of fertilizers.[54] Cotton has a virtually guaranteed market and enjoys access to credit facilities which food crops do not have. If small rural establishments, those most vulnerable to the drought, could be induced to participate more directly in the cash economy by the provision of such improved technology, backed up by credit and marketing facilities which make it profitable for the producer, then some significant progress might be made towards reducing the impact of the drought overall; that is, by providing the means whereby the low-income rural population may generate increased purchasing power, saving capacity and socio-economic independence.

Livestock production might be improved in drier areas, encouraging a certain amount of labour stabilization in *caatinga* which is at present thinly populated. Two major reasons for the extremely low productivity of animals in the *sertão* are, firstly, overgrazing and, secondly, lack of dry season forage. Overgrazing has harmful side-effects upon vegetation, soil fertility and on water available for consumption by forage species. The failure to store fodder leaves cattle vulnerable during the dry season, and especially during drought years when emergency feed is essential.[55]

Among the solutions which have been suggested for improving livestock production is more efficient use of rainfall and water resources, as well as more efficient use of fodder crops. Optimum use of the annual rainfall could be achieved by preventing run-off and improving vegetation. In the latter case it has been suggested that the replacement of herbaceous, annual species (which have short roots and dry up quickly) with deep-rooted perennials would increase the duration of the green season and provide fodder. Existing fodder resources could also be more efficiently used by storing roughage, limiting overgrazing and utilizing by-products to feed cattle.[56]

DNOCS has plans for regulating the supply of forage crops by combining large areas of drier *caatinga* with the agricultural calendar of its irrigation projects. The basic idea is that during winter (wet) months, project cattle would be grazed on the *caatinga*, while part of the irrigated land would produce roughage that would be stored and used to feed the animals during the summer months when dry pastures are inadequate.[57]

This combination of dry and wet land (*área sêca* and *área irrigada*) would enable fuller utilization of the capital resources invested in irrigation, with a ratio of one hectare of irrigated land to 20–100 hectares of dry pasture.[58] DNOCS hopes to incorporate 274,000 hectares of dry land to complement 100,000 hectares of irrigated land by 1980.[59] Similar suggestions have been made for improving *caatinga* pastures in areas of the interior untouched by irrigation, through improved water supply facilities by means of bore-holes and reservoirs.[60] To date, however, no areas of pastureland have been incorporated into irrigation projects. In theory such a strategy would provide additional employment, but it is difficult to estimate how much more in the absence of any concrete examples.

The latest government measure designed to promote agricultural development in the drought-affected zone is the so-called Projeto Sertanejo (*Programa Especial de Apoio ao Desenvolvimento da Região Semi-Árida do Nordeste*), officially inaugurated in August 1976. However, although the scheme is intended to improve agricultural technology and employment opportunities in selected areas of the *sertão* badly affected by drought, it seems unlikely that the small, weaker farmers most at risk will benefit. This is because resources are to be concentrated in existing DNOCS project areas. Credit and technical assistance will therefore be channelled into already privileged regions where the drought effects are least felt, the humid valleys. Furthermore, there is no provision for the redistribution of property either to small landowners or to landless labourers, those groups worst hit by the drought. There is every likelihood that funds made available under the Projeto Sertanejo will, like existing credit facilities, be monopolized by large landowners with political influence.

For the foreseeable future, therefore, it does not seem that either irrigation or any of the alternative strategies briefly discussed above will have a significant impact on countering the drought effects. Large sectors of the low-income rural population of the interior will remain vulnerable to even the smallest variation in the rainfall cycle because of their highly precarious, marginal economic positions. SUDENE, for example, still expects to play a key role in providing emergency relief to hundreds of thousands of *flagelados* on its work fronts. The agency's latest scheme plans the allocation of resources on work fronts and for other relief measures to accommodate 900,000 men in any future drought, compared with the 500,000 employed during the 1970 drought.[61]

The only other path open to those forced off the land is to migrate from the drought-afflicted areas altogether, either to towns within the North-East or further afield. In 1970 the net migratory loss for the North-East stood at 3,500,000, this figure having increased from 5% of the region's population in 1950 to 12% twenty years later.[62] Within the North-East, furthermore, 43% of migrants during the 1950–70 period were rural in origin.[63] The largest proportion of *nordestino* migrants went to São Paulo (39%) and Rio de Janeiro (23%), although Paraná and the Centre-West regions are becoming important.[64] Unfortunately there is no concrete, demographic evidence to show how strong migration from the *sertão* is compared with other regions in the North-East. However, one study suggests that emigration to the state of São Paulo, at least, may be stronger from the interior semi-arid zone than either the *agreste* or the *zona da mata*.[65]

Intra-regional migration within the North-East itself has been of paramount importance in the spontaneous reallocation of labour. From 1950–70 the 3,500,000 rural migrants to the towns accounted for 50% of urban growth.[66] During the period 1960–70, 20% of the region's population had made at least one move, and it appears that intra-regional migration is increasing at a faster rate than inter-regional migration.[67] Once again, very little data exist showing the importance of emigrants from the *sertão* in relation to movements within the North-East. Some indication of its intensity is given, nevertheless, by a study which shows that direct rural–urban migration in Fortaleza, Ceará, the state worst affected by droughts, reached 36% in the period 1960–70. This compares with 22% for Salvador and 14% for Recife.[68] The intensity of emigration from rural areas in the state of Ceará is attributable both to the changing agrarian structure which places more inhabitants in marginal positions, and to the 1970 drought which took its toll on the weakened population.

Tradition has it that the *sertanejo* always returns home from whichever part of Brazil he has emigrated to as soon as he hears that the winter rains have arrived in the *sertão*. In other words, seasonal migration appeared to predominate over permanent migration. Again, although very little information is available which could confirm or refute the popular image, there may be grounds for assuming that this trend has been reversed. A case study of the lower Tocantins region of Piauí, for example, noted the gradual replacement of temporary by permanent migration of rural workers.[69] If such a tendency were more general it would follow logically from the changes in the *sertão* rural structure discussed in chapter 2. It was noted that the fragmentation of small properties and expansion of *latifúndia* in the interior, both clearly observable phenomena, tend to displace labour, increase the economic marginality of the population affected and feed the rural–urban drift.

The droughts which periodically strike the *sertão*, then, dramatize deep-seated problems of rural weakness and the consequent vulnerability of its population. Drought in the North-East is not a simple climatic phenomenon which can be successfully tackled independently of this consideration. To say that the 'drought problem' is a total myth would be a gross exaggeration. However, it is true that many people have had, and continue to have, a vested interest in perpetuating the popular conception that the drought is the central cause of underdevelopment in the interior. Landowners and government planners anxious to promote water storage at the expense of every other development strategy use the drought as a convenient mask with which to obscure the basic causes of poverty.

Irrigation, as it has been implemented in North-East Brazil, does not benefit the drought-affected population of the interior and can have, at best, only a marginal impact in alleviating these hardships. This is a pity because irrigation currently absorbs two-thirds of government spending on agriculture in the North-East. That the emphasis on irrigation in rural development will continue seems fairly certain given the enormous commitments already made and those planned for the future. From the government's point of view it offers considerable advantages over the politically more controversial and troublesome areas of land redistribution, small farmer credit and cooperative formation. Irrigation schemes do not tamper significantly with the power structure, they create employment for thousands of bureaucrats and technicians and attract huge amounts of foreign money in search of development projects. In the rush to create green pastures, however, the rural poor appear to have been left by the wayside.

Notes

1. Drought in North-East Brazil – a general picture

1 GEIDA, *Programa Plurianual de Irrigação* (*PPI*), 13 vols. MINTER, Brasilia, 1971.
2 *Ibid.* vol. I, pp. 10–18, and Evandro de Souza Lima, *A Irrigação no Nordeste*, SUDENE, Fortaleza, 1973.
3 Joaquim Alves, *Historia das Sêcas*, Instituto do Ceará, Fortaleza, 1953, p. 20.
4 *Ibid.* p. 25.
5 Luis da Câmara Cascudo, *Geografia do Brasil Holandês*, José Olympio, R.J., 1956, p. 182.
6 Capistrano de Abreu, *Capítulos de Historia Colonial, 1500–1800*, Librarie Briguiet, R.J., pp. 187, 216–17.
7 Alves, *op. cit.* p. 32.
8 Antonil, cited by E. Silva Bruno, *Historia do Brasil, vol. VII*, Cultrix, São Paulo, 1967, pp. 84–5.
9 Spix and Martinius, cited by Alves, *op. cit.* p. 92.
10 *Ibid.* pp. 102–3.
11 George Gardner, *Travels in the Interior of Brazil, 1836–41*, London, 1846, p. 157.
12 Henry Koster, *Travels in Brazil*, London, 1816, p. 123.
13 Rodolfo Theóphilo, *História da Sêca no Ceará, 1877–80*, Imprensa Inglesa, R.J., 1922.
14 *Ibid.* p. 23.
15 *Ibid.* p. 28.
16 Celso Furtado, *Diagnosis of the Brazilian Crisis*, University of California Press, Berkeley and Los Angeles, 1965.
17 Theóphilo, *História da Sêca*, p. 417, and Rodolfo Theóphilo, *A Secca de 1915*, Imprensa Inglesa, R.J., 1922, p. 34.
18 Theóphilo, *História da Sêca*, p. 108.
19 *Ibid.* p. 109.
20 *Ibid.* pp. 141, 238–40.
21 Theóphilo, *A Secca de 1915*, pp. 14–15.
22 Alves, *op. cit.* p. 137.
23 DNOCS, *Polígono das Sêcas*, Fortaleza, 1968.
24 A.E. Hirschman, *Journeys Towards Progress*, 20th Century Fund, New York, 1963, pp. 18ff.
25 Arno S. Pearse, *Brazilian Cotton*, Manchester, 1923, ch. 3, 'Brazilian Public Works', pp. 58–9.
26 Hirschman, *op. cit.* p. 21.
27 Luis Carlos M. Pinheiro, 'Notas Sobre as Sêcas', *Boletim do DNOCS*, 20 (6), November 1959, p. 87.

28 Luis Carlos M. Pinheiro, 'Açudagem Pública', *Boletim do DNOCS*, 20 (6), November 1959.
29 DNOCS, *Açudes Públicos no Nordeste*, Fortaleza, 1972.
30 Raul Barbosa, *A Sêca de 1958*, BNB (ETENE), Fortaleza, 1958, p. 5, and Pinheiro, 'Notas Sobre as Sêcas', p. 125.
31 Pinheiro, 'Notas Sobre as Sêcas', p. 130.
32 Hirschman, *op. cit.* p. 70.
33 Antônio Callado, *As Industrias da Sêca e os 'Galileus' de Pernambuco*, Civilização Brasileira, R.J., 1960, p. 29.
34 *Ibid.* p. 12.
35 A.F. Montenegro, 'Tentativa de Interpretação das Eleições de 1958 no Ceará', *Revista Brasileira de Estudos Políticos*, April 1966, pp. 43–6.
36 Article 19 of the Inspectoria Statute of 20 February 1931, cited by Hirschman, *op. cit.* p. 76.
37 Alves, *op. cit.* pp. 137–8.
38 Hirschman, *op. cit.* p. 43.
39 Evaristo Leitão, *O Problema Social e Económico das Obras Contra a Sêca*, IFOCS, R.J., 1937, pp. 38–41.
40 J. Guimaraes Duque, *Solo e Água no Polígono das Sêcas*, DNOCS, Fortaleza, 1953, pp. 181–2.
41 These projects were Lima Campos, General Sampaio, Aires de Souza, São Gonçalo, Condado, Piranhas, Itans, Paulo de Brito Guerra, 'Evolução e Problemas da Irrigação no Nordeste', *Boletim do DNOCS*, 28 (2), July–December 1970.
42 *Ibid.* pp. 6–7. 'The population did not assimilate irrigation, they didn't understand it . . . it progressed very slowly.'
43 Hirschman, *op. cit.* pp. 72–8.
44 *Ibid.* pp. 82–3. Maranhão had just been included for planning purposes into the North-East, increasing the number of states from eight to nine. The governors' support for SUDENE was no doubt strengthened by DNOCS's abortive attempts in 1958 to influence voting against them.
45 DNOCS, *I Plano de Obras e Estudos, 1965–68*, Fortaleza, 1965, p. 26.
46 DNOCS, *Obras e Programas*, R.J., 1960.
47 SUDENE, *IV Plano Diretor de Desenvolvimento Econômico e Social do Nordeste, 1969–73*, Recife, 1968, pp. 281–90.
48 *Association pour L'Organisation de Missions de Coopération Technique* (ASMIC).
49 Grupo de Estudo de Base do Vale do Jaguaribe (GVJ), *Estudo Geral de Base do Vale do Jaguaribe*, 10 vols. SUDENE/ASMIC, Fortaleza, 1967.
50 Clovis Cavalcanti and Dirceu Pessoa. *Caráter e Efeitos da Sêca Nordestina de 1970*. SIRAC/SUDENE, Recife, 1973, pp. 173–5.
51 *Ibid.* p. 80 and DNOCS, *Frentes de Serviço*, Fortaleza, 1971, pp. 57–8.
52 Cavalcanti and Pessoa, *op. cit.* p. 76.
53 *Ibid.* p. 42.

54 Theóphilo, *História da Sêca*, pp. 99–102.
55 *Latin America* (newsletter), reports 22 May 1970 and 12 June 1970.
56 Emilio Garrastazú Medici, *Visão do Nordeste*, speech at SUDENE, Recife, 7 June 1970.
57 SUDENE, *IV Plano Diretor*, pp. 330ff.
58 Wilfred Alan Bacchus, *The Political Dimension of Water Resource Development: A Study of the Brazilian Northeast, 1964–71*, PhD thesis, University of Connecticut, 1973, p. 1094.
59 SEPLAN, *Programa de Desenvolvimento de Áreas Integradas do Nordeste (POLONORDESTE)*, Brasilia, 1974.
60 GEIDA, *PPI*, vol. I. pp. 16, 57 and table 4.8.
61 DNOCS, *Irrigação*, Fortaleza, 1975.
62 GEIDA, *PPI*, vol. I, pp. 6–9 and GEIDA, *Plano Nacional de Irrigação, Fase I*, Brasilia, 1970, pp. 1–20. For further information on goals see chapter 3 below.
63 SUDENE, *Relatório Sintético: Andamento do Programa de Irrigação no Nordeste, II Trimestre de 1974*, MINTER, Recife, 1974. Updated with additional information from DNOCS for 1974–5.
64 Djacir Meneses, *O Outro Nordeste*, José Olympio, R.J., 1937, p. 64.
65 See, for example, Graciliano Ramos, *Vidas Sêcas*, Martins, São Paulo, 1974, and Rachel de Queiroz, *O Quinze*, José Olympio, R.J., 1974.

2. O problema da sêca

1 The river systems are: São Francisco, Jaguaribe, Apodí, Piranhas–Açú, Parnaíba and Ceará Mirim. The *serras* or *chapadas* include: Ibiapaba, Baturité, Araripe, Martins Brejo, Teixeira, and Triunfo.
2 These cacti include the *xique-xique*, *facheiro*, *alastrado*, *mandacarú*, and *coroa-de-frade*. Euclides da Cunha, *Os Sertões*, R.J., 1968. pp. 32–9.
3 SEPLAN, *POLONORDESTE*, p. 9.
4 Manoel Correia de Andrade, 'A Terra e o Povo do Sertão', *Revista Geográfica*, 13, October 1975. L.F. Raposo Fontenelle, *Rotina e Fome em uma Região Cearense*, Imprensa Universitaria do Ceará, Fortaleza, 1969.
5 GVJ, *Estudo Geral. Apresentação e Síntese*, p. 33.
6 José Otamar de Carvalho, *Plano Integrado para o Combate Preventivo aos Efeitos das Sêcas no Nordeste*, MINTER, Brasilia, 1973, p. 22.
7 Cavalcanti and Pessoa, *Caráter e Efeitos da Sêca*, pp. 52–71.
8 SIRAC, *Diagnóstico e Potencialidades: Fase II, Vale do Jaguaribe, Vale do Piranhas–Açú*, and *Sertões Cearenses*, Recife, 1975.
9 *Diario de Pernambuco*, 24 April 1975.
10 *Diario de Pernambuco*, 18 April 1975.
11 *Diario de Pernambuco*, 25 and 28 April 1975.
12 'Sêca – Por Equanto um Aviso', *Veja*, 21 January 1976.
13 'Sêca – A Precária Sobrevivência', *Veja*, 4 February 1976.
14 The difference between a normal situation and a drought can be very fine in areas of the *sertão* where water supply facilities are very poor. During the dry season peasants are often forced to travel up to 50

kilometres to find another source if they are unable to pay for water brought in privately by lorry. This was the case in parts of western Pernambuco during October–November 1975, a non-drought year. See *Jornal do Brasil*, 7 November 1975.

15 *Diario de Pernambuco*, 22 June 1975.

16 Victor Du Bois, *The Drought in West Africa*, American Universities Field Staff Inc., West Africa Series, vol XV (1), General, 1974. Howard Brabyn, '6,000 Kilómetros da Africa Abandonados à Sêca', *Correio* (UNESCO), 3 (6), June 1975. Jacques Bugnicourt, 'Um Povo Privado do seu Meio Secular', *Correio* (UNESCO), 3 (6), June 1975.

17 Du Bois, *op. cit.*

18 USAID, *Desert Encroachment on Arable Lands: Significance, Causes and Control*, OST/AID, Washington D.C., 1972.

19 Luis Ricardo Leitão, 'Um Deserto no Coração do Brasil', *Manchete*, n.d., citing Vasconcelos Sobrinho.

20 *Diario de Pernambuco*, 22 June 1975.

21 'Project for the Modification of the North-Eastern Climate', announced by the Technical Aerospace Centre, São Paulo, *O Povo*, 8 October 1975.

22 Reports in *O Povo*, 8–9 October 1975.

23 DNOCS, *Programa e Obras*, Fortaleza, 1960, p. 3.

24 Cavalcanti and Pessoa, *op. cit.* p. 76.

25 *Ibid.* pp. 74–6.

26 Callado, *As Indústrias da Sêca.*

27 *O Globo*, 11 January 1976.

28 Alves, *História das Sêcas*, p. 95.

29 GTDN, *Uma Política de Desenvolvimento Econômico para o Nordeste*, 2nd ed. SUDENE, Recife, 1967, p. 67.

30 Cavalcanti and Pessoa, *op. cit.*

31 *Ibid.* p. 113.

32 SUDENE/IBRD, *Pesquisa do Tamanho Típico da Unidada de Produção Agrícola no Nordeste*, preliminary data (unpublished), Recife, 1975. The two organisations cooperated to produce this study.

33 Roger Lee Cunniff, *The Great Drought: Northeast Brazil, 1877–80*, PhD thesis, University of Texas at Austin, 1970, pp. 18–24.

34 Lopes de Andrade, *Introdução à Sociologia das Sêcas*, A. Noite, R.J., 1948, p. 164.

35 Cavalcanti and Pessoa, *op. cit.* p. 177.

36 *Ibid.* pp. 92–4, 212–13.

37 *Ibid.* pp. 191–5.

38 SUDENE/IBRD, *op. cit.*

39 *Ibid.*

40 Hidroservice, *Recuperação Hidroagrícola da Bacia de Irrigação de São Gonçalo*, 9 vols. São Paulo, 1971, vol. VI, *Condições Socio-Econômicas*, p. 1.3.2.

41 SIRAC, *Vale do Banabuiú: Zona de Transição*, DNOCS, 1973, p. 50.

42 SIRAC, *Baixo Jaguaribe: Estudo de Viabilidade*, vol. II, *Dados Básicos*, part 4, *Situação Atual*, DNOCS, Recife, 1974, p. 52.

43 Sir Richard Burton, *The Highlands of Brazil*, 2 vols. London, 1869, vol II, p. 432.
44 Orville Derby, *Reconhecimento Geológico do Vale do São Francisco*, Typografia Nacional, R.J., 1880, p. 23.
45 Data from my own field research in Ceará, Brazil, 1975.
46 Victor R. Cadaxa, *Rural Trade Unionism and Politics in Northeastern Brazil, 1955–64*, MA thesis, London University, 1973, p. 10.
47 Cunniff, *op. cit.* pp. 28–31.
48 Thomaz Pompeu de Souza Brasil, *Ensaios Estatísticos da Província do Ceará*, 2 vols. Fortaleza, 1867, vol. II, p. 393.
49 Cunniff, *op. cit.* p. 96.
50 Peter Singleman, 'Political Structures and Social Banditry in Northeast Brazil', *Journal of Latin American Studies*, 7 (1), 1974.
51 M.V. Vilaça and R.C. de Albuquerque, *Coronel, Coronéis*, Tempo Brasileiro, R.J., 1965, pp. 38–9.
52 *Ibid.*
53 B.J. Chandler, *The Feitosas and the Sertão dos Inhamuns*, University of Florida Press, Gainsville, 1972.
54 Singleman, *op. cit.* p. 77.
55 Cadaxa, *op. cit.* p. 10 and Maria Auxiliadora Ferraz de Sá, *Dos Velhos aos Novos Coronéis*, PIMES, Recife, 1974, pp. 23–5.
56 Cavalcanti and Pessoa, *op. cit.* p. 139.
57 George F. Patrick and J.J. de Carvalho Filho, *Grupos de Pobreza na Agricultura Brasileira: Relatório Preliminar*, FIPE, São Paulo, 1975, p. 28.
58 IBGE, *Agricultural Census*, 1970.
59 *Ibid.*
60 Hidroservice, *op. cit.* vol. VI.
61 A.W. Johnson, *Sharecroppers of the Sertão*, Stanford University Press, California, 1971.
62 Steven S. Cheung, *The Theory of Share Tenancy*, University of Chicago Press, Chicago and London, 1969.
63 Johnson, *op. cit.* pp. 31–40.
64 *Ibid.* pp. 40–1.
65 *Ibid.* p. 38 and A.W. Johnson and B. Siegel, 'Wages and Income in Ceará, Brazil', *Southwestern Journal of Anthropology*, 25 (1), Spring 1969, pp. 4–6.
66 Hidroservice, *op. cit.* vol. VI, and DNOCS, *Estudo Socio-Econômico de uma Área Rural do Vale do Acaraú*, Fortaleza, 1973, pp. 76–80, and W.H. Nicholls and Ruy Miller Paiva, *Noventa e Nove Sítios e Fazendas: Estrutura e Produtividade da Agricultura Brasileira, 1963, III A Região do Carirí no Ceará: Crato*, Fundação Getulio Vargas, Graduate Centre for Latin American Studies, Vanderbilt University, R.J., 1968, p. 105.
67 Vicenzo d'Apote, *Bases para uma Política de Colonização e Reforma Agrária no Nordeste do Brasil*, SUDENE, Recife, 1972, pp. 115–17.
68 Of 435 contracts surveyed in the Sertão do Pajeú, only five (1%) were written and the rest verbal. SIRAC, *Estudo da Rentabilidade dos Estabelecimentos Agropecuários e das Condições do Arrenda-*

mento da Terra no Estado de Pernambuco, SUDENE/DAA, Recife, 1974, table 4.2.

69 In the Acaraú valley, Ceará, wage-labourers (locally known as *alugados*) are almost totally unemployed for six months of the year, employment opportunities being limited by the high degree of family labour used. Their general economic situation is more desperate as they do not grow their own food. DNOCS, *Estudo Socio-Econômico*, p. 101.

70 W.H. Nicholls and Ruy Miller Paiva, 'The Structure and Productivity of Brazilian Agriculture', *Journal of Farm Economics*, 47 (2), May 1965, p. 358.

71 Yony de Sá Barreto Sampaio, 'Modelos de Otimização na Agricultura no Nordeste do Brasil, Com Ênfase em Agricultura do Consorciação em Parceria', *Communicações*, 9. PIMES, Refice, 1975.

72 Patrick and Carvalho Filho, *op. cit.* p. 37.

73 Nicholls and Paiva, *Noventa e Nove Fazendas, III*, p. 106.

74 Cavalcanti and Pessoa, *op. cit.* p. 191, and Patrick and Carvalho Filho, *op. cit.* pp. 33–4.

75 SIRAC, *Vale do Banabuiú*, p. 100.

76 DNOCS, *Estudo Socio-Econômico*, p. 64.

77 SIRAC, *Baixo Jaguaribe*, vol. II, part 4, p. 67.

78 Nicholls and Paiva, *Noventa e Nove Fazendas, III*, pp. 102–3.

79 Johnson and Siegel, *op. cit.* pp. 8–9.

80 D. Gross and B.A. Underwood, 'Technological Change and Caloric Costs: Sisal Agriculture in Northeastern Brazil', *American Anthropologist*, 73, 1971.

81 Charles Slater et al., *Market Processes in the Recife Area of Northeast Brazil,* Michigan University, East Lansing, 1969.

82 BNB, *Mercado e Commercialização do Algodão no Nordeste.* Fortaleza, 1964, p. 101.

83 *Ibid.* pp. 103–4, and DNOCS, *Vale do Fidalgo: Estudo de Pre-Viabilidade,* vol. I, part 1, Teresina, 1973.

84 BNB, *Mercado do Algodão*, p. 101, and DNOCS, *Vale do Fidalgo*, vol. I, part 2, p. 566.

85 BNB, *Mercado do Algodão*, p. 104, and Hidroservice, *op. cit.* vol. VI, p. 2.4.8.

86 SUDENE, 'Estudos Preliminares do Setor Agrícola do Nordeste', *Boletim Econômico*, 1 (1), January–June 1969, p. 66.

87 SUDENE, 'Estudos Preliminares', p. 66.

88 Cavalcanti and Pessoa, *op. cit.* pp. 151–3.

89 *Ibid.* pp. 211–12.

90 Emergency Credit for the North-East. CEN was designed to subsidize employment-creating investment on the *fazendas* in an attempt to stem the flow of unemployed to the work fronts. At the same time it would result in permanent capital formation, in contrast to the fronts which were considered by many to be a waste of public money in the long run. The cost of CEN was small (Cr$20 million compared with a Cr$332 million expenditure on work fronts) but it was claimed to have created 56,000 jobs against the fronts' 450,000,

a very impressive ratio. Funds were disbursed only upon receipt of proof of wage payment to his workers by the borrower and a commitment to keep them on the land. Mailson Ferreira da Nóbrega, 'Desenvolvimento da Agro-pecuária Nordestina: O Banco do Brasil e Ação Governamental', *Boletim do Banco do Brasil*, 7 (3), 1973.

91 George Patrick, *Desenvolvimento Agrícola no Nordeste*, IPEA/INPES, R.J., 1972, pp. 36, 45, 53 and IBGE, *Anuário Estatístico do Brasil*, 1971, pp. 148, 161–2, 165.

92 Report in *O Globo*, 11 January 1976.

93 Cavalcanti and Pessoa, *op. cit.* pp. 211–12.

94 Based on information from the Ministry of Agriculture.

95 *Programa de Redistribuição de Terras e Estímulos à Agro-Indústria do Nordeste* (Programme of Land Redistribution and Stimuli to North-Eastern Agro-Industry), instituted by Decree-Law 1179 of 6 July 1972.

96 Based on data from the Bank of Brazil – 2nd Region.

97 *Ibid.*

98 Patrick, *Desenvolvimento Agrícola*, pp. 43ff.

99 SUDENE, 'Estudos Preliminares', p. 80.

100 Francisco Sá Jnr, 'O Desenvolvimento da Agricultura Nordestina e a Função das Atividades de Subsistência', *Estudos CEBRAP*, 3 January 1973.

101 S. Foreman and J.F. Reigelhaupt, 'Market Place and Marketing System: Towards a Theory of Peasant Economic Integration', *Comparative Studies in History and Society*, 12 (12), 1970.

102 Patrick, *Desenvolvimento Agrícola*, pp. 169–70.

103 *Ibid.* pp. 171–2.

104 IBGE Census data suffer from the disadvantage of using administrative rather than juridical landholding units but also have advantages. Firstly, data can be compared over time. Secondly, they may be broken down into distinct geographical areas (*litoral*, *agreste*, *sertão*). Thirdly, rural 'establishment' rather than the strictly legal 'property' division is conceivably more useful to indicate an economic unit of effective labour absorption. No other source combines these features.

105 For the whole North-East the rate of *minifúndio* fragmentation has been slower (including *agreste* and *litoral* regions). The average size of establishments up to 10 hectares has fallen from 3.90 hectares in 1940 to 2.99 hectares in 1970. The rate of concentration of larger holdings, on the other hand, has been similar, with those over 100 hectares falling from 10.6% to 8% of the total number (1940–60) but maintaining their share of cultivated land at just over 74%. Data compiled from SUDENE, 'Estudos Preliminares', table 20, and IBGE, *Agricultural Census*, 1970.

106 SUDENE, *Projetos Agro-Industriais e Agro-Pecuários Aprovados: Situação até Junho de 1974*, Recife, 1974.

107 SUDENE, 'Estudos Preliminares', p. 15.

108 Sá Jnr, *op. cit.* The SUDENE/IBRD survey confirms the IBGE statistics, pointing out that family labour forms the largest single occupational category. The same study also shows, moreover, that

family labour is overwhelmingly concentrated on *minifúndios* of less than 10 hectares. IBGE also documents this fact: in 1960 58% of family labour in the North-East was concentrated on holdings of under 10 hectares. SUDENE, 'Estudos Preliminares', table 30.

109 Figures for the *sertão* show a comparable rate of increase with the North-East generally. Family labour as a proportion of the total in the whole region went up from 64% in 1960 to 83% in 1970. SUDENE, 'Estudos Preliminares', tables 29 and 30, and IBGE, *Agricultural Census*, 1970.

110 IBGE, *Agricultural Census*, 1970.

3. The origins and state of irrigation in North-East Brazil

1 See for example DNOCS, *Brumado River Valley Irrigation Project*, Fortaleza, n.d.

2 Guerra, 'Evolução e Problemas', p. 7.

3 *Ibid.* p. 7.

4 Hidroservice, *Recuperação Hidroagrícola*, vol. VI, *Condições Socio-Econômicas*, p. 5.3.1.

5 *Ibid.* p. 2.3.7.

6 *Ibid.* table 3.3.3.

7 A Franco-Brazilian mission made a detailed study of the Jaguaribe valley from 1962–5, and shortly afterwards work on the Morada Nova project was started by SUDENE, although it was later taken over by DNOCS. See GVJ, *Estudo Geral.* In the São Francisco valley the Bebedouro Project was commenced in 1969 but remained small scale with just over one hundred families installed by 1974. SUDENE, *Relatório Sintético*.

8 GEIDA, *PPI*, vol. I, table 2.4, p. 18.

9 *Ibid.* pp. 8–15 and GEIDA, *PPI*, vol. III, annex 8, p. 15.

10 GEIDA, *PPI*, vol. I, p. 8.

11 GEIDA, *PPI*, vol. III, annex 8, p. 12.

12 *Ibid.* p. 18.

13 GEIDA, *PPI*, vol. I, pp. 6–9, and GEIDA, *Plano Nacional*, p. B1–7.

14 GEIDA, *PPI*, vol. I, table 4.11.

15 *Ibid.* table 3.5, p. 33.

16 *Ibid.* p. 9.

17 GEIDA, *Plano Nacional*, p. B1–6.

18 GEIDA, *PPI*, vol. III, annex 8, p. 18.

19 *Ibid.* p. 18.

20 GEIDA, *PPI*, vol. I, table 4.8.

21 DNOCS, *Irrigação*.

22 *Ibid.*

23 SIRAC, *Projeto de Desenvolvimento Rural Integrado do Nordeste (Áreas Prioritárias)*, IPEA, 1973.

24 Riorden Roett, *The Politics of Foreign Aid in the Brazilian Northeast*, Vanderbilt University Press, Nashville, 1972, p. 149.

25 SIRAC, *Projeto de Desenvolvimento*, pp. 16–17.

26 DNOCS, *Irrigação*.

27 DNOCS, *Programa de Irrigação, 1971–74*, Fortaleza, 1971.
28 SUDENE, *Relatório Sintético*. Updated with data from DNOCS.
29 GEIDA, *PPI*, vol. XII, annex 32.
30 Pinheiro, 'Açudagem Pública'.
31 DNOCS, *Projeto Sumé*, Recife, 1968, p. 16.
32 *Ibid*. p. 35.
33 Data from São Gonçalo meteorological station.
34 DNOCS, *I Plano de Obras e Estudos*.
35 Hidroservice, *Recuperação Hidroagrícola, Relatório Final (Carta de Apresentação)*.
36 Distributed land is still currently the property of DNOCS although it is being debated by the agency if, and how, the land can be handed over to the farmers (either as a gift or bought).
37 GVJ, *op. cit.*
38 DNOCS, *Jaguaribe: O Vale e o DNOCS*, Fortaleza, 1975. These projects are: Morada Nova, Baixo Jaguaribe, Icó–Lima Campos, Várzea do Boi, Quixabinha, Santo Antônio de Russas, Ema and Riacho do Sangue.

4. Employment and stabilization of the rural population

1 William R. Cline, 'Análise de Custo-Benefício de Projetos de Irrigação no Nordeste', *Pesquisa e Planejamento Econômico*, 2 (2), December 1972, p. 259.
2 Calculated from d'Apote, *Bases para uma Política*, pp. 115–18.
3 DNOCS, *Irrigação*.
4 Cline, 'Análise de Custo-Benefício', p. 259.
5 7.9 billion *cruzeiros* or US$1.3 million in 1971. SUDENE, *Produto e Formação Bruta de Capital, 1965–71*, Recife, 1973, p. 18.
6 Lynn H. Distelhorst, *Difusão do Espírito Empresarial na Extensão Rural: Avaliação do Sucesso nos Primeiros Perímetros Irrigados do Sertão Nordeste Brasileiro, 1972–3*, DNOCS, Fortaleza, 1973.
7 See, for example, GEIDA, *PPI*, vol. XIII, *Características Básicas dos Projetos Avaliados.*
8 SUCAM, *Levantamento Demográfico e Habitacional de São Gonçalo*, Ministerio de Saúde, João Pessoa, August 1971, p. 2.
9 Hidroservice, *Recuperação Hidroagrícola*, vol. IX, *Viabilidade Técnico-Econômico, Financeira e Administrativa da Recuperação Hidroagrícola, Alternativa*, p. 108.
10 DNOCS, *Projeto Sumé*, p. 17.
11 SIRAC survey data, as well as estimates from project engineers and social workers.
12 Interview with head of Expropriation Sector, DNOCS 2nd Regional Directorate, Fortaleza, 30 September 1975, and SIRAC, *Vale do Banabuiú*.
13 Faculdade de Filosofia Dom Aureliano Matos (Limoeiro do Norte), *Projeto Morada Nova–Banabuiú: Levantamento Sócio-Situacional*, DNOCS, Fortaleza, 1973, p. 19.

14 Interview with head of Expropriation Sector, DNOCS 2nd Regional Directorate, Fortaleza, 30 September 1975.
15 DNOCS, *Plano Anual de Exploração, 1975*, São Gonçalo, 1975.
16 Interview with head of Expropriation Sector, DNOCS 2nd Regional Directorate, Fortaleza, 30 September 1975.
17 Hidroservice, *op. cit.*, vol. VI, *Condições Socio-Econômicas*, table 2.2.2.
18 DNOCS, *Seleção de Familias*, mimeo, 1973.
19 DNOCS, *Normas Provisórias para a Seleção de Famílias de Irrigantes*, mimeo, n.d.
20 *Ibid.*
21 *Ibid.*
22 Tenant farmers who rent a small plot of land alongside the reservoir and are thus totally protected from the drought, able to cultivate at any time of the year, as well as doing some fishing in the *açude.*
23 Hidroservice, *op. cit.*, vol. VI, p. 43.
24 *Ibid.* tables 3.4.1 and 3.4.2.
25 SIRAC, *Vale do Banabuiú*, tables 33–4.
26 Fundo de Assistência para o Trabalhador Rural, set up by Decree no. 69919 of 11 January 1972.
27 I.e. he must have a *Carteira de Trabalho e Previdência Social.*
28 GEIDA, *PPI*, vol. I, pp. 6–9, and GEIDA, *Plano Nacional*, p. B1.7.
29 José Lins Albuquerque, *Sobre Irrigação em Regiões Subdesenvolvidas*, DNOCS, Fortaleza, 1971, pp. 6–7.
30 IJNPS, *Minuta do Ante-Projeto para o Planejamento Sociológico dos Projetos Gurgueia e Fidalgo*, mimeo, Recife, January 1975.
31 Interview with José Otamar de Carvalho, head of Research Division, Ministry of the Interior, Brasilia, 19 March 1975.

5. Rural incomes

1 GEIDA, *PPI*, vol. I, tables 4.11, 3.5, pp. 9 and 33.
2 GEIDA, *PPI*, vol. III, annex 8, p. 18.
3 Albuquerque, *Sobre Irrigação*, p. 14.
4 Hidroservice, *Recuperação Hidroagrícola, Relatório Final (Carta de Apresentação).*
5 SUDENE/SCET, *Vale do Rio Banabuiú: Anti-Projeto do Valorização Hidroagrícola. Relatório Geral – Texto*, SUDENE, Recife, 1969, p. 269.
6 SCET, *Açude Público Várzea do Boi: Projeto de Irrigação*, vol. I, DNOCS, Fortaleza, 1972, p. 10.
7 GEIDA, *PPI*, vol. I, pp. 6–9 and GEIDA, *Plano Nacional*, p. B1.7.
8 SCET, *Açude Público Sto. Antônio de Russas: Projeto de Irrigação*, vol. I, DNOCS, Fortaleza, 1972, p. 91.
9 Income data taken from DNOCS, *Cooperativa dos Irrigantes da Paraíba Ltda, Balanço de Encerramento Realizado em 31.12.74*, São Gonçalo, 1975.
10 DNOCS, *CIVAB-Demonstrativo de Receitas e Despesas (1974);*

Prestação de Contas, Morada Nova, 1975. CIVAB is the *Cooperativa dos Irrigantes do Vale do Banabuiú.*

11 DNOCS, *Balanço dos Colonos*, Sumé, 1974.

12 DNOCS, *Cooperativa dos Irrigantes, Balanço.*

13 DNOCS, *CIVAB: Prestação de Contas.*

14 DNOCS, *Balanço dos Colonos.*

15 DNOCS, *Dívidas dos Colonos à CIVAB (1973–74)*, Morada Nova, 1975.

16 Distelhorst, *Difusão do Espírito Empresarial.*

17 *Ibid.*

18 *Ibid.* p. 166.

19 William R. Cline, *Irrigation Projects in Brazil's Northeast: Case Studies in Social Cost-Benefit Analysis*, IPEA/INPES, R.J., 1971, revised 1972. Unpublished.

20 *Ibid.*

21 Jorge Coelho, *Considerações em Torno do Programa de Agricultura Irrigada na Zona Semi-Árida do Nordeste*, Centro Estadual de Treinamento para o Nordeste, Recife, 1975.

22 Cline, 'Análise de Custo-Benefício', p. 258.

23 Colin Clark, *The Economics of Irrigation*, 2nd ed. Pergamon Press, Oxford, 1970, pp. 62–71.

24 Cline, *Irrigation Projects in Brazil's Northeast*, p. 21.

25 DNOCS, *Plano Anual de Exploração, 1975.*

26 DNOCS, *Projeto de Irrigação Morada Nova: Plano de Exploração para o Ano de 1975*, Morada Nova, 1974.

27 Bacchus, *The Political Dimension of Water Resource Development*, p. 902.

28 *Ibid.* p. 935.

29 *Ibid.* p. 717.

30 Interview with project engineer, São Gonçalo, 24 January 1975.

31 Coelho, *Considerações em Torno do Programa*, p. 13.

32 Distelhorst, *op. cit.* p. 168.

33 Hidroservice, *op. cit. Relatório Final (Carta de Apresentação).*

34 Hidroservice, *op. cit.* vol. IX, *Viabilidade Técnico-Econômico, Financeria e Administrativa da Recuperação Hidroagrícola, Alternativa*, p. 85.

35 *Ibid.* p. 86.

36 DNOCS, *Cooperativa dos Irrigantes, Balanço*, and DNOCS, *Plano Anual de Exploração.*

37 Interview with project engineer, São Gonçalo, 24 January 1975.

38 SUDENE/SCET, *Vale do Rio Banabuiú.*

39 Interview with IRYDA (Spanish Ministry of Agriculture) agronomist on contract to DNOCS, Morada Nova, 26 September 1975.

40 DNOCS, *Plano Anual de Exploração.*

41 DNOCS, *Plano de Exploração Agronômica*, Sumé, 1974.

42 DNOCS, *Projeto de Irrigação Morada Nova*, and DNOCS, *Balancete de Verificação dos Saldos, Julho de 1975*, Morada Nova, 1975.

43 Distelhorst, *op. cit.*

44 *Ibid.* p. 184.

45 *Ibid.* p. 185.
46 *Ibid.* p. 183. Parental response as an indicator of future trends should, however, be treated with caution.
47 Interview with social worker, São Gonçalo, 24 February 1975.
48 Bacchus, *op. cit.* pp. 717 and 932.
49 Distelhorst, *op. cit.* pp. 206–7.
50 *Ibid.* p. 220.
51 Interview with social worker and *práticos*, São Gonçalo, 28 February 1975.
52 See for example DNOCS, *Discriminação das Parcelas de Crédito para Mão de Obra*, São Gonçalo, 1975.

6. Agricultural production

1 GEIDA, *PPI*, vol. I, p. 20 and vol. II, annex 3, p. 1.
2 GEIDA, *PPI*, vol. I, p. 20 and vol. II, annexes 3 and 4.
3 GEIDA, *PPI*, vol. II, annex 3, pp. 2–3.
4 GEIDA, *PPI*, vol. I, p. 20.
5 *Ibid.* p. 25. See table 18 below.
6 GEIDA, *PPI*, vol. II, annex 3, p. 4.
7 GEIDA, *PPI*, vol. I, graph 4.11.
8 *Ibid.* graphs 4.11 and 4.12.
9 GEIDA, *PPI*, vol. V, annex 27, p. 19.
10 GEIDA, *PPI*, vol. I, pp. 31–2.
11 *Ibid.*
12 *Ibid.* table 3.5, p. 33.
13 GEIDA, *PPI*, vol. II, annex 7, table 1, p. 9. In Fortaleza it was found that intermediaries made 70–80% of the profit on perishable crops and 30% on cereals. See SUDENE, *Aspectos da Commercialização de Produtos Horti-Fruti-Granjeiros e Cereais na Cidade de Fortaleza*, DAA, Fortaleza, 1973, and SUDENE, *Aspectos da Commercialização de Produtos Horti-Fruti-Granjeiros na Cidade do Recife*, Recife, 1973. Similar trends were observed in Recife. See SUDENE, *Commercialização de Alimentos na Cidade do Recife*, Recife, 1972.
14 By the setting up of CIDAs (*Companhias Integradas de Desenvolvimento Agrícola*). See chapter 7 for fuller details.
15 GEIDA, *PPI*, vol. I, pp. 43–5.
16 DNOCS, *Relatório Anual: Pesquisas Agropecuárias*, Recife, 1974.
17 Coelho, *Considerações em Torno do Programa*, p. 38. SUDENE, *Relatório Sintético*.
18 DNOCS, *Plano Anual de Exploração*.
19 DNOCS, *Balancete de Verificação*.
20 Interviews with administration and cooperative personnel, Sumé, December 1974.
21 BNB, *Perspectivas de Desenvolvimento do Nordeste até 1980: Síntese*, Fortaleza, 1973, pp. 49–50.
22 *Ibid.* p. 50.
23 Hidroservice, *Recuperação Hidroagrícola, Relatório Final (Carta de Apresentação)*, p. 10.

24 SUDENE/SCET, *Vale do Rio Banabuiú*, p. 222.
25 Clovis Cavalcanti, *Aspectos da Commercialização de Alimentos na Cidade do Recife*, SUDENE/ANCARPE, Recife, 1971.
26 Hidroservice, *op. cit. Relatório Final (Carta de Apresentação)*, p. 10.
27 GEIDA, *PPI*, vol. I, pp. 43–6.
28 DNOCS, *Seleção de Agricultores: Fichas para Classificação de Candidatos*, Fortaleza, 1975.
29 Interviews with social workers, São Gonçalo and Morada Nova, February–September 1975.
30 DNOCS, *Estatuto da Cooperativa dos Irrigantes do Vale do Banabuiú*, Morada Nova, 1972. DNOCS, *Estatuto da Cooperativa dos Irrigantes da Paraíba*, São Gonçalo, 1973.
31 DNOCS, *Relatório de Atividades do Serviço de Promoção Social e Cooperativista*, Recife, 1974, p. 12.

7. Alternatives to irrigation

1 'A Situação Criada pela Implantação do Projeto de Irrigação São Desidério', *Cadernos do CEAS*, 44, pp. 46, 51.
2 'Irrigação, Uma Solução para o Desemprêgo', *Visão*, 22 December 1975, pp. 68–73.
3 Bacchus, *The Political Dimension of Water Resource Development*, p. 718.
4 Colin Clark, *Starvation or Plenty?*, Secker and Warburg, London, 1970, p. 117. Clark's observations concern irrigation based on the family plot system rather than collective means of agricultural production.
5 Arthur Gaitskell, *Gezira: A Story of Development in the Sudan*, Faber and Faber, London, 1959.
6 R.J. Harrison-Church, 'Observations on Large-Scale Irrigation Development in Africa', *Agricultural Economics Bulletin for Africa (FAO)*, 4, November 1963, p. 35.
7 Georges Brausch, 'Change and Continuity in the Gezira Region of the Sudan', *International Social Science Journal*, 16 (3), 1964, p. 335.
8 David Barkin, *Los Beneficiarios del Desarrollo Regional*, Sep/Setentas, Mexico City, 1972, pp. 160–76. Irrigation was started here by Italian immigrants earlier in the century, but in 1938 the Mexican government expropriated the valley lands and collectivized production under the *ejido* system. From 1950–70 the government expanded the irrigated area by six times and built an extensive infrastructure of transport facilities and social services. Large numbers of private farmers were attracted to the region. The original inhabitants of the valley, the *ejidatarios*, could only secure a limited amount of credit to finance cotton production from official sources, the *Banco Nacional de Credito Ejidatal* (BNCE), because they only had a limited amount of collateral to offer as security. Private landowners, on the other hand, expanded production rapidly, acquiring all the credit they needed from private banks for sowing and for setting up cotton gins to extract the maximum amount of profit. The *ejidatarios* eventually

found that they were better off renting out their lands to private producers, although this was illegal, rather than cultivating cotton themselves. The BNCE built a small processing plant in 1956 and has increased the supply of credit to *ejidos* but the private sector has continued to dominate the cotton industry.

9 Ernest Feder, *The New Penetration of the Agricultures of the Underdeveloped Countries by the Industrial Nations and Their Multi-National Concerns*, mimeo, Institute of Social Studies, The Hague, 1976.

10 Eugene A. Havens and William L. Flinn, *Green Revolution Technology and Community Development: the Limits of Action Programs*, mimeo, n.d.

11 T.J. Byres, 'The Dialectics of India's Green Revolution', *South Asian Review*, 5, 1971–2, pp. 99–116.

12 Harrison-Church, *op. cit.* pp. 41–3.

13 Henry Orenstein, 'Notes on the Ecology of Irrigation Agriculture in Contemporary Peasant Societies', *American Anthropologist*, 67, 6 (1), p. 1529.

14 Burton Pasternak, 'Social Consequences of Equalizing Irrigation Access', *Human Organization*, 27 (4), 1968.

15 T.S. Epstein, *Economic Development and Social Change in South India*, Manchester University Press, 1962.

16 See for example, Helio Galvão, *O Mutirão no Nordeste*, Serviço de Informação Agrícola, R.J., 1959.

17 Benno Galjart, 'Old Patrons and New', *Sociologia Ruralis*, 7 (4), 1967.

18 Charles Erasmus, 'Community Development and the Encojido Syndrome', *Human Organization*, 27 (1), 1968.

19 Gaitskell, *op. cit.* and Harrison-Church, *op. cit.* p. 18.

20 Harrison-Church, *op. cit.* p. 19.

21 Gaitskell, *op. cit.* p. 43.

22 Harrison-Church, *op. cit.* p. 11.

23 *Ibid.* p. 12.

24 For example, the *fazendas* Catalunha, Milano and others in the valley. See Bacchus, *op. cit.* p. 937–45 and *Visão*, 22 December 1975.

25 Arthur D. Little Montor, *Projeto de Desenvolvimento de Agroindústria no Nordeste do Brasil*, IPLAN, Brasilia, 1974.

26 SIRAC, *Diagnóstico e Potencialidades, Vale do Moxotó* and *Vale do Piranhas–Açú.*

27 SIRAC, *Diagnóstico e Potencialidades, Vale do Ceará Mirim*, p. 26.

28 George F. Patrick, 'Efeitos de Programas Alternativas do Govêrno sobre a Agricultura do Nordeste', *Pesquisa e Planejamento Econômico*, 4, February 1974, pp. 61–5.

29 Carl H. Gotsch, 'Credit Programs to Reach Small Farmers', *Development Digest*, 12 (1), January 1974.

30 See chapter 2, and Johnson, *Sharecroppers of the Sertão.*

31 Sutti Ortiz, 'The "Human Factor" in Social Planning in Latin America' in R. Apthorpe, ed., *People Planning and Development Studies*, Frank

Cass & Co, London, 1970, and Sutti Ortiz, 'Reflections on the Concept of "Peasant Culture" and Peasant Cognitive Systems', in T. Shanin (ed.), *Peasants and Peasant Societies*, Penguin, Harmondsworth, Middx, 1971.

32 Ronald Tinnemeir, 'Technology, Profit and Agricultural Credit', *Development Digest*, 12 (1), January 1974.

33 At the regional level the *Companhia de Abastecimento do Nordeste* (CANESA) was created by SUDENE but the only unit in operation is the *Central de Abastecimento do Recife* (CEASA). Similar projects for Salvador and Fortaleza were never implemented. Certain national organizations operate in the North-East such as CIBRASEN, COBAL, SUNAB and the Bank of Brazil, which participates in the government programme of minimum prices as an agent for *Commisão de Financiamento de Produção* (CFP). See Manoel Figueroa, *Cuestiones de Política Agrária Regional: Nordeste do Brasil*, SUDENE/DAA, Recife, 1973, p. 203.

34 *Ibid.* pp. 200–4.

35 *Ibid.* pp. 195–225.

36 *Ibid.* p. 206.

37 Ginaldo Cavalcanti, 'Commercialização Agrícola', *Boletim de Agricultura*, SUDENE, Recife, January–June 1974, p. 94.

38 *Ibid.* pp. 95–6.

39 Figueroa, *op. cit.* p. 209.

40 Patrick, 'Efeitos de Programas Alternativas', pp. 76–80.

41 *Ibid.* p. 79.

42 *Ibid.* p. 79.

43 Jorge Coelho, mimeo, untitled, SUDENE/DAA, Recife, 1975.

44 Helvecio Dornas, *O Processo da Reforma Agrária na Região Nordeste*, mimeo, SUDENE/DAA, Recife, 1975.

45 Coelho, mimeo, untitled, and SUDENE, *Projetos Agro-Industriais*.

46 Dornas, *op. cit.*

47 GTDN, *Uma Política de Desenvolvimento*, p. 82.

48 d'Apote, *Bases para Uma Política*, pp. 115–18.

49 *Ibid.* pp. 86, 98. The author calculated that 3,800,000 hectares are available for colonization in the states of Bahia, Piauí and Maranhão, which could absorb 259,000 *sertanejos*.

50 The costs of colonization have been put at US$2,547 per family, against US$12,300–15,300 for irrigation. See Vania P. Tavares et al, *Colonização Dirigida no Brasil*, IPEA/INPES, R.J., 1972, pp. 114–16. Another estimate gives figures of Cr$17,400 and Cr$104,400 respectively. See Coelho, *Considerações em Torno do Programa*, p. 29.

51 *Jornal do Brasil*, July 1968.

52 W.H. Nicholls, 'The Brazilian Agricultural Economy', In R. Roett (ed.), *Brazil in the Sixties*, Vanderbilt University Press, Nashville, 1972, p. 175.

53 *Ibid.* p. 174.

54 Ruy Miller Paiva, *Apreciação Preliminar dos Problemas da Agricultura Brasileira*, UN (CEPLA), R.J., 1953.

55 P. Guilbride, *FAO/IPEANE Project for Livestock Development in the North-East: Objectives and Implementation*, FAO, Recife, 1973.
56 *Ibid.* pp. 6–7.
57 Interview with Dirceu Pessoa, Director of SIRAC, the major outside consultants to DNOCS in this field.
58 SIRAC, *Projeto de Desenvolvimento*, pp. 26–7.
59 DNOCS, *Irrigação*.
60 SIRAC, *Diagnóstico e Potencialidades, Sertões Cearenses*.
61 SUDENE, *Plano de Ação para Emergencia Contra as Calamidades de Sêca e Enchentes, 1973*, Recife, 1972, p. 46.
62 Helio A. de Moura, 'As Variações Migratórias no Nordeste, 1940–70', *Revista Econômica*, 4 (14), October–December 1972, p. 23.
63 Milton da Mata et al., *Migrações Internas no Brasil: Aspectos Econômicos e Demográficos*, IPEA/INPES, R.J., 1973, p. 78.
64 Moura, 'As Variações Migratórias', p. 26.
65 Manoel de Souza Barros, *Éxodo e Fixação*, Serviço de Informação Agrícola, R.J., 1957, p. 39. The author gives a coefficient per 100 emigrants from Pernambuco to São Paulo from 1941–50 of 2.95 (*sertão*), 1.84 (*agreste*) and 0.45 (*zona da mata*).
66 Helio A. de Moura, *Migrações para as Grandes Cidades no Nordeste: Intensidade e Características Demográficas*, mimeo, MINTER, R.J., 1974, p. 4.
67 *Ibid.* pp. 50, 55, 156.
68 *Ibid.* pp. 3–4.
69 E.N. Gonzales and M.I. Bastos, *Migração Rural e o Trabalho Volante na Agricultura Brasileira*, mimeo, University of Brasilia, presented at *Encontro Brasileiro de Estudos Populacionais*, R.J., 1974.

Bibliography

Abreu, Capistrano de. *Capítulos de História Colonial, 1500–1800*, Livraria Briguiet, R.J., 1954.

Albuquerque, José Lins. *Sobre Irrigação em Regiões Subdesenvolvidas*, DNOCS, Fortaleza, 1971.

Alves, Joaquim. *História das Sêcas*, Instituto do Ceará, Fortaleza, 1953.

Andrade, Lopes de. *Introdução à Sociologia das Sêcas*, A. Noite, R.J., 1948.

Andrade, Manoel Correia de. *A Terra e o Homem no Nordeste*, Brasiliense, São Paulo, 1964.

'A Terra e o Povo do Sertão', *Revista Geográfica*, 13, October 1975.

Apote, Vicenzo d'. *Bases para uma Política de Colonização e Reforma Agrária no Nordeste do Brasil*, SUDENE, Recife, 1972.

Araujo, Genesio M. de. *Contribuição Passada e Presente do DNOCS à Solução dos Problemas Nordestinos*, DNOCS, Fortaleza, 1969.

Bacchus, Wilfred Alan. *The Political Dimension of Water Resource Development: A Study of the Brazilian Northeast, 1964–71*, PhD thesis, University of Connecticut, 1973.

Banco do Nordeste do Brasil (BNB). *Mercado e Commercialização do Algodão no Nordeste*, Fortaleza, 1964.

Perspectivas de Desenvolvimento do Nordeste até 1980: Síntese, Fortaleza, 1973.

Suprimento de Gêneros Alimentícios para a Cidade de Fortaleza, Fortaleza, 1964.

Barbosa, Raul. *A Sêca de 1958*, BNB (ETENE), Fortaleza, 1958.

Barkin, David. *Los Beneficios del Desarrollo Regional*, Sep/Setentas, Mexico City, 1972.

Barros, Manoel de Souza. *Éxodo e Fixação*, Serviço de Informação Agrícola, R.J., 1957.

Brabyn, Howard. '6,000 Kilómetros da Africa Abandonados à Sêca', *Correio* (UNESCO), 3 (6), June 1975.

Brasil, Thomaz Pompeu de Souza. *Ensaios Estatísticos da Província do Ceará*, 2 vols. Fortaleza, 1867.

Brausch, Georges. 'Change and Continuity in the Gezira Region of the Sudan', *International Social Science Journal*, 16 (3), 1964.

Bruno, E. Silva. *História do Brasil, vol. VII*, Cultrix, São Paulo, 1967.

Bugnicourt, Jacques. 'Um Povo Privado do seu Meio Secular', *Correio* (UNESCO), 3 (6), June 1975.

Burton, Sir Richard. *The Highlands of Brazil*, 2 vols. London, 1869.

Byres, T.J. 'The Dialectics of India's Green Revolution', *South Asian Review*, 5, 1971–2.

Cadaxa, Victor R. *Rural Trade Unionism and Politics in Northeastern Brazil, 1955–64*, MA thesis, London University, 1973.

Cadernos do CEAS. 'A Situação Criada pela Implantação do Projeto de Irrigação São Desidério', 44, 1976.

Callado, Antônio. *As Indústrias da Sêca e os 'Galileus' de Pernambuco*, Civilização Brasileira, R.J., 1960.

Carvalho, José Otamar de. *Plano Integrado para o Combate Preventivo aos Efeitos das Sêcas no Nordeste*, MINTER, Brasilia, 1973.

Cascudo, Luis da Câmara. *Geografia do Brasil Holandês*, José Olympio, R.J., 1956.

Cavalcanti, Clovis. *Aspectos da Commercialização de Alimentos na Cidade do Recife*, SUDENE/ANCARPE, Recife, 1971.

Cavalcanti, Clovis and Pessoa, Dirceu. *Caráter e Efeitos da Sêca Nordestina de 1970*, SIRAC/SUDENE, Recife, 1973.

Cavalcanti, Ginaldo. 'Commercialização Agrícola', *Boletim de Agricultura*, SUDENE, Recife, January–June 1974.

Cheung, Steven S. *The Theory of Share Tenancy*, University of Chicago Press, Chicago and London, 1969.

Chandler, B.J. *The Feitosas and the Sertão dos Inhamuns*, University of Florida Press, Gainsville, 1972.

Clark, Colin. *The Economics of Irrigation*, 2nd ed. Pergamon Press, Oxford, 1970.

Starvation or Plenty? Secker & Warburg, London, 1970.

Cline, William R. 'Análise de Custo-Benefício de Projetos de Irrigação no Nordeste', *Pesquisa e Planejamento Econômico*, 2 (2), December 1972.

Irrigation Projects in Brazil's Northeast: Case Studies in Social Cost-Benefit Analysis, IPEA/INPES, R.J., 1971, revised, 1972. Unpublished.

Coelho, Jorge. *Considerações em Torno do Programa de Agricultura Irrigada na Zona Semi-Árida do Nordeste*, Centro Estadual de Treinamento para o Nordeste, Recife, 1975.

PROTERRA: Relatório Analítico, SUDENE, Recife, 1974.

Mimeo, untitled, SUDENE/DAA, Recife, 1975.

Cunha, Euclides da. *Os Sertões*, R.J., 1968.

Cunniff, Roger Lee. *The Great Drought: Northeast Brazil, 1877–80*, PhD thesis, University of Texas at Austin, 1970.

Departamento Nacional de Obras Contra as Sêcas (DNOCS). *Açudes Públicos no Nordeste*, Fortaleza, 1972.

Balancete de Verificação dos Saldos, Julho de 1975, Morada Nova, 1975.

Balanço dos Colonos, Sumé, 1974.

Brumado River Valley Irrigation Project, Fortaleza, n.d.

CIVAB-Demonstrativo de Receitas e Despesas (1974): Prestação de Contas, Morada Nova, 1975.

Cooperativa dos Irrigantes da Paraíba Ltda, Balanço de Encerramento Realizado em 31.12.74, São Gonçalo, 1975.

Discriminação das Parcelas de Crédito para Mão de Obra, São Gonçalo, 1975.

Dívidas dos Colonos à CIVAB (1973–74), Morada Nova, 1975.

Estatuto da Cooperativa dos Irrigantes da Paraíba, São Gonçalo, 1973.

Estatuto da Cooperativa dos Irrigantes do Vale do Banabuiú, Morada Nova, 1972.
Estudo Socio-Econômico de uma Área Rural do Vale do Acaraú, Fortaleza, 1972.
Frentes de Serviço, Fortaleza, 1971.
Irrigação, Fortaleza, 1975.
Jaguaribe: O Vale e o DNOCS, Fortaleza, 1975.
Normas Provisórias para a Seleção de Familias de Irrigantes, mimeo, n.d.
Obras e Programas, R.J., 1960.
Plano Anual de Exploração, 1975, São Gonçalo, 1975.
Plano de Exploração Agronômica, Sumé, 1974.
Polígono das Sêcas, Fortaleza, 1968.
I Plano de Obras e Estudos, 1965–68, Fortaleza, 1965.
Projeto de Irrigação Morada Nova: Plano de Exploração para o Ano de 1975, Morada Nova, 1974.
Projeto Sumé, Recife, 1968.
Programa de Irrigação, 1971–74, Fortaleza, 1971.
Programa e Obras, Fortaleza, 1960.
Relatório Anual: Pesquisas Agropecuárias, Recife, 1974.
Relatório de Atividades do Serviço de Promoção Social e Cooperativista, Recife, 1974.
Seleção de Agricultores: Fichas para Classificação de Candidatos, Fortaleza, 1975.
Seleção de Famílias, mimeo, 1973.
Vale do Fidalgo; Estudo de Pre-Viabilidade, Teresina, 1973.
Derby, Orville. *Reconhecimento Geológico do Vale do São Francisco*, Typografia Nacional, R.J., 1880.
Diario de Pernambuco. Reports, 18, 24, 25, 28 April 1975 and 22 June 1975.
Distelhorst, Lynn H. *Difusão do Espírito Empresarial na Extensão Rural: Avaliação do Sucesso nos Primeiros Perímetros Irrigados do Sertão Nordeste Brasileiro, 1972–3*, DNOCS, Fortaleza, 1973.
Dornas, Helvécio. *O Processo da Reforma Agrária na Região Nordeste*, mimeo, SUDENE/DAA, Recife, 1975.
Du Bois, Victor. *The Drought in West Africa*, American Universities Field Staff Inc., West Africa Series, vol. XV (1), General, 1974.
Duque, J. Guimaraes. *Solo e Água no Polígono das Sêcas*, DNOCS, Fortaleza, 1953.
Epstein, T.S. *Economic Development and Social Change in South India*, Manchester University Press, 1962.
Erasmus, Charles. 'Community Development and the Encojido Syndrome', *Human Organization*, 27 (1), 1968.
Faculdade de Filosofia Dom Aureliano Matos, (Limoeiro do Norte). *Projeto Morada Nova–Banabuiú: Levantamento Sôcio-Situacional*, DNOCS, Fortaleza, 1973.
Feder, Ernest. *The New Penetration of the Agricultures of the Underdeveloped Countries by the Industrial Nations and Their Multinational Concerns*, mimeo, Institute of Social Studies, The Hague, 1976.

Ferraz de Sá, Maria Auxiliadora. *Dos Velhos aos Novos Coroneis*, PIMES, Recife, 1974.

Figueroa, Manoel. *Cuestiones de Política Agrária Regional: Nordeste do Brasil*, SUDENE/DAA, Recife, 1973.

Fontenelle, L.F. Raposo. *Rotina e Fome em uma Região Cearense*, Imprensa Universitaria do Ceará, Fortaleza, 1969.

Foreman S. and Reigelhaupt, J.R. 'Market Place and Marketing System: Towards a Theory of Peasant Economic Integration', *Comparative Studies in History and Society*, 12 (12), 1970.

Furtado, Celso. *Diagnosis of the Brazilian Crisis*, University of California Press, Berkeley and Los Angeles, 1965.

Gaitskell, Arthur. *Gezira: A Story of Development in the Sudan*, Faber & Faber, London, 1959.

Galjart, Benno. 'Old Patrons and New', *Sociologia Ruralis*, 7 (4), 1967.

Galvão, Helio. *O Mutirão no Nordeste*, Serviço de Informação Agrícola, R.J., 1959.

Gardner, George. *Travels in the Interior of Brazil, 1836–41*, London, 1846.

O Globo. Report, 11 January 1976.

Gonzales, E.N. and Bastos, M.I. *Migração Rural e o Trabalho Volante na Agricultura Brasileira*, mimeo, University of Brasilia, presented at *Encontro Brasileiro de Estudos Populacionais*, R.J., 1974.

Goodman, David and Cavalcanti, Roberto. *Incentivos à Industrialização e Desenvolvimento do Nordeste*, IPEA/INPES, R.J., 1974.

Gotsch, Carl H. 'Credit Programs to Reach Small Farmers', *Development Digest*, 12 (1), January 1974.

Gross, D. and Underwood, B.A. 'Technological Change and Caloric Costs: Sisal Agriculture in North-eastern Brazil', *American Anthropologist*, 73, 1971.

Grupo de Estudo de Base do Vale do Jaguaribe (GVJ). *Estudo Geral de Base do Vale do Jaguaribe*, 10 vols. SUDENE/ASMIC, Fortaleza, 1967.

Grupo Executivo de Irrigação para o Desenvolvimento Agrícola (GEIDA). *Plano Nacional de Irrigação, Fase I*, Brasilia, 1970.
Programa Plurianual de Irrigação (PPI), 13 vols. MINTER, Brasilia, 1971.

Grupo de Trabalho para o Desenvolvimento do Nordeste (GTDN). *Uma Política de Desenvolvimento Econômico para o Nordeste*, 2nd ed. SUDENE, Recife, 1967.

Guerra, Paulo de Brito. 'Evolução e Problemas da Irrigação no Nordeste', *Boletim do DNOCS*, 28 (2), July–December 1970.

Guilbride, P. *FAO/IPEANE Project for Livestock Development in the North-East: Objectives and Implementation*, FAO, Recife, 1973.

Harrison-Church, R.J. 'Observations on Large-Scale Irrigation Development in Africa', *Agricultural Economics Bulletin for Africa (FAO)*, 4, November 1963.

Havens, Eugene A. and Flinn, William L. *Green Revolution Technology and Community Development: the Limits of Action Programs*, mimeo, n.d.

Hidroservice (Engenharia de Projectos Ltda). *Recuperação Hidroagrícola da Bacia de Irrigação de São Gonçalo*, 9 vols. São Paulo, 1971.

Hirschman, A.O. *Journeys Towards Progress*, 20th Century Fund, New York, 1963.

Instituto Brasileiro de Geografia e Estatística (IBGE). *Agricultural Census*, 1940, 1950, 1960, 1970.

Anuário Estatístico do Brasil, 1971.

Instituto Joaquim Nabuco de Pesquisas Socais (IJNPS). *Minuta do Ante-Projeto para o Planejamento Sociológico dos Projetos Gurguiea e Fidalgo*, mimeo, Recife, January 1975.

Instituto de Planejamento Econômico e Social (IPEA). *Agricultura Brasileira: Comportamento Passado, Situação Atual e Perspectivas de Crescimento*, Brasilia, 1973.

Johnson, A.W. *Sharecroppers of the Sertão*, Stanford University Press, California University Press, California, 1971.

Johnson, A.W. and Siegel, B. 'Wages and Income in Ceará, Brazil', *Southwestern Journal of Anthropology*, 25 (1), Spring 1969.

Jornal do Brasil. Reports, July 1968 and 7 November 1975.

Koster, Henry. *Travels in Brazil*, London, 1816.

Latin America (newsletter). Reports, 17 May 1974, 22 May 1970, 12 June 1970.

Leitão, Evaristo. *O Problema Social e Econômico das Obras Contra a Sêca*, IFOCS, R.J., 1937.

Leitão, Luis Ricardo. 'Um Deserto no Coração do Brazil', *Manchete*, n.d.

Lima, Evandro de Souza. *A Irrigação no Nordeste*, SUDENE, Fortaleza, 1973.

Lisboa, Miguel Arrojado. 'O Problema das Sêcas', *Boletim do DNOCS*, 20 (6), November 1959.

Mata, Milton da (et al.). *Migrações Internas no Brasil: Aspectos Econômicos e Demográficos*, IPEA/INPES, R.J., 1973.

Medici, Emilio Garrastazú. *Visão do Nordeste*, speech at SUDENE, Recife, 7 June 1970.

Meneses, Djacir. *O Outro Nordeste*, José Olympio, R.J., 1937.

Montenegro, A.F. 'Tentativa de Interpretação das Eleições de 1958 no Ceará', *Revista Brasileira de Estudos Políticos*, April 1966.

Montor, Arthur D. Little. *Projeto de Desenvolvimento de Agroindustria no Nordeste do Brasil*, IPLAN, Brasilia, 1974.

Moura, Helio A. de. 'As Variações Migratórias no Nordeste, 1940–70', *Revista Econômica*, 4 (14), October–December 1972.

Migrações para as Grandes Cidades do Nordeste: Intensidade e Características Demográficas, MINTER, R.J., 1974.

Nicholls, W.H. 'The Brazilian Agricultural Economy', in R. Roett (ed.), *Brazil in the Sixties*, Vanderbilt University Press, Nashville, 1972.

Nicholls, W.H. and Paiva, Ruy Miller. *Noventa e Nove Sítios e Fazendas: Estrutura e Produtividade da Agricultura Brasileira, 1963, III A Região do Cariri no Ceará: Crato*, Fundação Getulio Vargas, Graduate Centre for Latin American Studies, Vanderbilt University, R.J., 1968.

'The Structure and Productivity of Brazilian Agriculture', *Journal of Farm Economics*, 47 (2), May 1965.

Nóbrega, Mailson Ferreira de. 'Desenvolvimento da Agro-pecuária Nordestina: O Banco do Brasil e Ação Governmental', *Boletim do Banco do Brasil*, 7 (3), 1973.

Orenstein, Henry. 'Notes on the Ecology of Irrigation Agriculture in Contemporary Peasant Societies', *American Anthropologist*, 67, 6 (1).

Ortiz, Sutti. 'The "Human Factor" in Social Planning in Latin America', in R. Apthorpe (ed.), *People Planning and Development Studies*, Frank Cass & Co., London, 1970.

'Reflections on the Concept of "Peasant Culture" and Peasant Cognitive Systems', in T. Shanin (ed.), *Peasants and Peasant Societies*, Penguin, Harmondsworth, Middx, 1971.

Paiva, Ruy Miller. *Apreciação Preliminar dos Problemas da Agricultura Brasileira*, UN (CEPLA), R.J., 1953.

Pasternak, Burton. 'Social Consequences of Equalizing Irrigation Access', *Human Organization*, 27 (4), 1968.

Patrick, George F. *Desenvolvimento Agrícola no Nordeste*, IPEA/INPES, R.J., 1972.

'Efeitos de Programas Alternativas do Govêrno sobre a Agricultura do Nordeste', *Pesquisa e Planejamento Econômico*, 4, February 1974.

Patrick, George and Carvalho Filho, J.J. de. *Grupos de Pobreza na Agricultura Brasileira: Relatório Preliminar*, FIPE, São Paulo, 1975.

Pearse, Arno S. *Brazilian Cotton*, Manchester, 1923.

Pinheiro, Luis Carlos M. 'Açudagem Públîca', *Boletim do DNOCS*, 20 (6), November 1959.

'Notas Sobre as Sêcas', *Boletim do DNOCS*, November 1959.

'Obras Contra os Efeitos das Sêcas (Sêca de 1958)', *Boletim do DNOCS*, 3 (9), February 1959.

O Povo. Reports, 8, 9 October 1975.

Queiroz, Rachel de. *O Quinze*, José Olympio, R.J., 1974.

Ramos, Graciliano. *Vidas Sêcas*, Martins, São Paulo, 1974.

Roett, Riorden. *The Politics of Foreign Aid in the Brazilian Northeast*, Vanderbilt University Press, Nashville, 1972.

Sá Jnr, Francisco. 'O Desenvolvimento da Agricultura Nordestina e a Função das Atividades de Subsistência', *Estudos CEBRAP*, 3, January 1973.

Sales Pinto, Dr F. Letter to head of SUCAM, 18 January 1973.

Sampaio, Yony de Sá Barreto. 'Modelos de Otimização na Agricultura no Nordeste do Brasil, Com Ênfase em Agricultura de Consorciação em Parceria', *Communicações*, 9, PIMES, Recife, 1975.

Secretaria de Planejamento (SEPLAN). *Programa de Desenvolvimento de Áreas Integradas do Nordeste (POLONORDESTE)*, Brasilia, 1974.

Serviços Integrados de Assesoria e Consultoria (SIRAC). *Baixo Jaguaribe: Estudo de Viabilidade*, DNOCS, Recife, 1974.

Diagnóstico e Potencialidades: Fase II, Sertões Cearenses, Vale do Piranhas–Açú, Vale do Moxotó, Vale do Ceará Mirim, Vale do Jaguaribe, Recife, 1975.

Estudo da Rentabilidade dos Estabelecimentos Agropecuários e das Condições do Arrendamento da Terra no Estado de Pernambuco, SUDENE/DAA, Recife, 1974.

Projeto de Desenvolvimento Rural Integrado do Nordeste (Áreas Prioritárias), IPEA, Recife, 1973.

Singleman, Peter. 'Political Structure and Social Banditry in Northeast Brazil', *Journal of Latin American Studies*, 7 (1), 1976.

Slater, Charles (et al.). *Market Processes in the Recife Area of Northeast Brazil*, Michigan University, East Lansing, 1969.

Societé Central pour l'Equipment du Territoire-Coopération (SCET). *Açude Público Sto. Antônio de Russas: Projeto de Irrigação*, vol. I, DNOCS, Fortaleza, 1972.

Açude Público Várzea do Boi: Projeto de Irrigação, vol. I, DNOCS, Fortaleza, 1972.

Superintendência de Campanhas de Saúde Pública-Setor Paraíba (SUCAM). *Levantamento Demográfico e Habitacional de São Gonçalo*, Ministerio de Saúde, João Pessoa, August 1971.

Superintendência do Desenvolvimento do Nordeste (SUDENE). *Aspectos da Commercialização de Produtos Horti-Fruti-Granjeiros e Cereais na Cidade de Fortaleza*, DAA, Fortaleza, 1973.

Aspectos da Commercialização de Produtos Horti-Fruti-Granjeiros e Cereais na Cidade do Recife, Recife, 1973.

Commercialização de Alimentos na Cidade do Recife, Recife, 1972.

'Estudos Preliminares do Setor Agrícola do Nordeste', *Boletim Econômico*, 1 (1), January–June 1969.

Perdas na Commercialização de Produtos Horti-Fruti-Granjeiros na Cidade de Fortaleza, Fortaleza, 1973.

Plano de Ação para Emergência Contra as Calamidades de Sêca e Enchentes, 1973, Recife, 1972.

Produto e Formação Bruta de Capital, 1965–71, Recife, 1973.

Projetos Agro-Industriais e Agro-Pecuários Aprovados: Situação até Junho de 1974, Recife, 1974.

IV Plano Diretor de Desenvolvimento Econômico e Social do Nordeste, 1969–73, Recife, 1968.

Relatório Sintético: Andamento do Programa de Irrigação no Nordeste, II Trimestre de 1974, MINTER, Recife, 1974.

SUDENE/SCET. *Vale do Rio Banabuiú: Anti-Projeto de Valorização Hidroagrícola. Relatório Geral – Texto*, SUDENE, Recife, 1969.

SUDENE/IBRD. *Pesquisa do Tamanho Típico da Unidade de Produção Agrícola no Nordeste*, preliminary data (unpublished), Recife, 1975.

Tavares, Vania P. (et al.). *Colonização Dirigida no Brasil*, IPEA/INPES, R.J., 1972.

Theóphilo, Rodolfo. *A Secca de 1915*, Imprensa Inglesa, R.J., 1922.

História da Sêca no Ceará, 1877–80, Imprensa Inglesa, R.J., 1922.

Tinnemeir, Ronald. 'Technology, Profit and Agricultural Credit', *Development Digest*, 12 (1), January 1974.

United States Agency for International Aid and Development (USAID). *Desert Encroachment on Arable Lands: Significance, Causes and Control*, OST/AID, Washington D.C., 1972.

Veja. 'Sêca – A Precária Sobrevivência', report 4 February 1976.
'Sêca – Por Enquanto um Aviso', report 21 January 1976.
Vilaça, M.V. and Albuquerque, R.C. de. *Coronel, Coronéis*, Tempo Brasileiro, R.J., 1965.
Visão. 'Irrigação, Uma Solução para o Desemprêgo', 22 December 1975.
'Um Projeto para o Nordeste', 22 December 1975.

Index